TALKING TO THE WORLD

TALKING TO THE WORLD

John Newgas

CENTURY COMMUNICATIONS
LONDON

First published in Great Britain in 1984
by Century Communications Ltd,
12–13 Greek Street,
London W1V 5LE

ISBN 0 7126 0558 4

Printed in Great Britain in 1984 by
Hazell, Watson & Viney Ltd. (Member of the BPCC Group)
Aylesbury, Bucks.

To Celia, who patiently sat through
the writing of this book.

Contents

Introduction

Hacking – Your Neighbour is Just a Phone Call Away.

Like most people I bought my first home computer not for work but for fun. I killed Invaders, chased aliens and explored Dungeons and Caverns, but soon found that staying alone in my room playing games didn't really amuse me, and put away my computer for a long time, hardly ever touching it. Then I discovered 'Hacking'.

'Hacking' is the enthusiasts' term for calling and speaking to other people and their computers using the phone. This was everything that I had really hoped to find. You never know who will be there at the other end, or what may be found. Nearly always there is a distant and interested keyboard with similar interests awaiting you.

Many calls like this caused my interest to grow and, after some time, I started to organize a message board of my own. This expanded and through it many people have had a chance to 'speak' to me. Finally, a friend approached me and asked me to write this book to help others to follow the same path.

While some people might want to use their computer to keep their accounts or for other types of administration, this book is about a very different kind of computing - not inside the computer but out in the real world. By reaching this world your computer stops being a games playing machine and becomes a channel and assistant. So put away your Space Invader cartridges, sweep up the bloodstains from Adventure Dungeons and see what can be done.

In this world you can meet and talk to other users, not face to face but screen to screen; huge games can be played, both against computers, and against several human players at once. Magazines can be read and, if you feel bold enough, written and edited.

This is the world of communications. By its nature it is a social world. While you might be alone in a room, your relationship is now not with your machine but with other users, sometimes in countries across the world.

This whole field is new for private users. Professionals have been working in it for many years, but only now is it becoming available and affordable for the amateur enthusiast. As this is a new area, you will be able to pioneer new methods and ideas. This book is a guide book to the trails that are available.

The area of communications is expanding at a fast pace. What was once an exotic speciality is now available from High Street shops all over the country with the coming of Prestel. The home microcomputer has changed from being a rare and expensive piece of equipment to a common thing, almost at a throw-away price. The possible uses are only just beginning to catch up with the technical features available.

I don't know whether we will all be working from our homes in a few years' time with computers using 'Communications'. The techniques and costs are practical – but it is for all of us to try these techniques out and see which of them we really want. No longer do we need to be told by advisers what we may do. This book is meant not just as a guide to the technology but to the uses which are here today.

Just knowing what can be done is not enough; the technical details have also to be mastered. As well as a guide to the landmarks of possibilities, a do-it-yourself handbook is needed. Because this book is intended to fulfil both requirements, the mixture of chapters may seem unusual. To avoid excessive use of jargon you will find chapters alternate between those on general issues and those covering the more meaty technical points. The chapters are not always directly connected, but the pattern is generally kept to. You may never want to read some of the more technical chapters – they could be ignored altogether.

However like most areas of computing, communications and hacking are so enjoyable as to be almost addictive and, as you make more use of the available facilities, you will find that there are many things in the technical chapters which can help.

If you are one of the people who like to skip through a book and just pick out the parts which seem interesting, please just go straight ahead. Why read about things which you are not interested in ? If you buy this book rather than browse on bookstore shelf then the information you need will be ready here for whenever you need it. If you are a bookstore browser (shame on you) then try starting with Chapter 3 – 'A Call to a Bulletin Board' to see what all this is about. This book is written for your enjoyment. Few things are more exciting than calling in to a message service and being told that there is a message waiting for you – it's just like being given a present to unwrap! More enjoyable still is organizing and running a message service for other people and meeting, in a distant way, other users.

CHAPTER 1
Overview

My dictionary tells me that communications means the exchange of information or emotions between two or more beings. This definition was written long before computers existed, but still explains the purpose of this book. Some examples will help to make the possibilities clearer.

Nearly everyone is familiar with a community notice board where people put up small advertisements or notices. Bulletins and letters may be seen on school or church notice boards, and even outside many newsagents' shops. At first, boards may only display advertisements of items for sale:

'One used accordion for Sale – No Sensible Offers refused – Contact Michael at 213-9562 or leave a message inside.'

Perhaps another note might ask all who are interested in starting a course to sign their names below. Soon a good notice board can become a centre for solving problems and even meeting others.

There is no need for our notice board to be in any special place. It is how it is used which makes it important. The act of writing on or reading a notice board is something that everyone can do. Since today everyone uses the telephone and many people use microcomputers, it is natural for the notice and bulletin boards to use these modern techniques.

A telephone is simply used to call up a central bulletin board kept on a computer. This is not very different from the original idea. The telephone just saves the journey to where the board is usually kept. If the board is kept on a computer then it would be normal to look at it on a display screen or television set. It doesn't matter which screen or television you look at, the adverts and notices will still be the same.

To call from your home and use a microcomputer is an easy process. Some specialized equipment is needed but this is generally readily available.

The actual technical side of what happens is simple. The telephone normally only sends sounds, not letters, so an adaptor has to be fitted to convert the letters into sounds at one end, and from sounds back to letters for the computer at the other end. This adaptor is called a modem. Because there are two ends to conversation it is normal to call the larger system (in this example the bulletin board) the host, while you, the caller in the home, are are usually known as the terminal.

With computers the technical means often seems to overwhelm the real purpose. This is the greatest problem in books such as this. Words like *modem* will be used all the way through this book. These abbreviated terms will help you to understand the technical chapters. All the terms will be explained in more detail later.

To put a letter such as D on a screen or printer most computers use a simple code based on only zero and one. Nearly all microcomputers (and many larger machines) have found that only eight pieces of information (or 'bits') need to be used for each letter. Usually all eight bits will be sent at the same time over eight separate wires inside the computer or even to a printer. When using a phone line only one piece of information at a time can be sent; so each bit is sent one at a time in turn from the computer. This is called serial transmission.

To make different types of computers and electrical accessories compatible when using this serial method, a common plug and wiring standard were agreed some years ago. This is now known as the RS-232 Standard. Nearly all microcomputers and modems use it today.

Making a computer call seems complex, but really it is just a chain of little steps. When you press a key on the keyboard the computer sets eight wires to hold the code for that particular letter. The computer then sends the information from each wire out, one at a time, to a converter unit which makes sure that the voltage going out is to the standard. The little chain of pulses is converted by the modem to musical tones which are then sent down the telephone. All this is done without the user knowing about it, usually in under a thirtieth of a second for each letter.

Once two computers can talk to each other and exchange letters and information then they become typing telephones and can be used to exchange gossip, receive news or even transfer programs without having to use cassettes or floppy disks.

As well as casual notice boards, there are many professional systems available to help with various jobs in many fields. There are large libraries of information (often called data-banks) available for little more than the price of a telephone call.

This book mainly concentrates on public bulletin boards. This is because they contain the four main services available for remote communications – public notice boards, electronic mail, information banks and games. The notice board concept has been described above; the other facilities can be looked on as specialized forms of these.

Electronic mail, the computer equivalent of a telegram sent to a personal post box, is rapidly becoming a cheap and effective way of sending mail, especially when letters have to be sent urgently to other countries. This should not be confused with its forerunner, Telex. The new electronic mail service leaves your letters in a pigeonhole waiting for you to collect them from a central point, whereas Telex is simply a telephone/typewriter which must always be ready and available for other users.

The computer may have a large store of information of some kind. Many people use the computer communication method to take advantage of this, creating a home reference library. This method goes further in that much serious research is now done this way, by calling into a stored bank of information. This has become such an important area that now many companies exist only to service that need for information. Examples include Prestel in Britain or Dialog, a giant series of encyclopaedic indexes, information banks and notes in the United States.

Playing games may seem a trivial use of such technology but, solitary amusements are never much fun. There is an increasing popularity for games that can be played from one's home and yet still involve other people. Most users still get a great deal of pleasure from single-player games played on large machines offering more features than they normally have available at home.

In many countries specialized communications facilities, similar to the telephone network, are available for computer use at lower than normal telephone costs. These should not confuse you. The route is not important, but the end result is. While different countries offer different commercial products the free public bulletin board is almost universal. The wide variety of facilities all over the world have various advantages and disadvantages for the private user. Many excellent private services welcome private users without charge. Some other commercial services offer very economic rates for private subscribers, and will welcome your subscription and use of their services.

These, and other similar possibilities, are developing at a rapid pace. The same facilities could be used to allow office work to be done from the home, or to allow greatly improved leisure facilities. No-one can foretell the future directions of growth, yet clearly there is rapid development in this area. By the time you have read this book new ideas will be starting their life, hopefully you, too, will have found something that you would like to do. It is only by trying things out that proficiency and experience are won.

CHAPTER 2

The Scope of the Book

Hacking and communications encompass such a large area that it is impossible to explain everything in a single book but, for nearly everyone, communications means using the telephone and so this is the main method described here. This does not involve any extraordinary kind of telephone, but just the usual line found nearly everywhere. In some countries there are special low-cost long-distance telephone routes connected to the public phone system. These are very useful if you can get access to them.

The special digital communications links which are mainly used commercially will not be explained here. In Britain they are called Packet Switch Stream, or PSS for short. They can sometimes be reached through the ordinary dial-up phone system and so are not completely out of reach of amateur hackers. This world of PSS networks is worth exploring because of the huge number of large libraries and services it provides but, unfortunately, nearly all of these charge fairly steep rates to users. The main difference from the telephone-based systems is that these digital networks send the information, not as changing tones over a telephone designed for voice, but as special electrical signals very close to those used inside your own computer.

The professional world originally used the communications methods which hackers use now. As business pressures have become more intense and mainframe computers grown larger, so the professional mainframe users have moved to much higher flow rates of exchanging information. Amateurs and hackers normally send text at about 30 letters each second. This is about as fast as one can read the text. Prestel and other similar systems send at about 120 letters per second. This is much faster than one can read, and fills up a page on a screen in only six or seven seconds. The professional user might start at a speed of 240 letters per second and use any speed up to about 2,000 letters a second.

These high-speed methods are rarely used by amateurs and hackers. They are used between large business computers and usually require special on wiring to transfer the information. However, the standards and speeds used by hackers and others shouldn't be belittled. These are much faster than Telex – in fact, much faster than anyone can type, and generally quite reliable.

The same methods are used (often with the same equipment as well) for commercial electronic mail. This is really just a commercial version of the bulletin board service. The main difference between amateur and commercial message services is that the amateur ones can usually only answer one caller at a time while the professional boards may be able to answer several, perhaps even hundreds, simultaneously.

The technical details and methods are identical to those in commercial use. There are agreed international standards which are the same for amateur and professional alike. This means that the equipment and methods you use as an amateur are the same as you are likely to use for serious work.

This common approach means that you will be able to call and talk to some of the more interesting serious systems all over the world. Those which readily accept calls from outside casual users and private callers, such as Compuserve in the United States and some of the smaller systems, are well worth a phone call. Other systems such as the specialized data-bases holding information on various special subjects won't be covered here. They are normally used by those who need their services – and if you need them then you will probably know a great deal about the particular one you want.

While anyone can use electronic and computer equipment mechanically without any understanding, this never gives good results. The key to success in overcoming problems when making calls is an understanding of what is happening and why. A large part of the technical chapters is devoted to helping you with these hidden stages.

So that different types of computer can connect and speak to each other there are agreements on linking up. Many standards are used, some official and others which are not really standards at all but just agreements between a few people. To join in worldwide communications you must stick to normal standards, which cover everything from plugs and connectors, and the voltages used to connect to the telephone system at the mechanical end, to how letters and text are sent at the computer software end. All the standards are really important and many of them have chapters to themselves. The main standards that you need to know about is the RS-232, which is used for connecting computers to accessories like modems, and the standard that the modems themselves use.

The RS-232 standard is for transmitting information in a stream, with one wire in each direction. This means that only one bit of information can be sent at a time. There are many variations on this; Chapters 6 and 12 explain the likely settings in use. The high-speed transmissions also often use a similar serial method but they save a little time by having synchronized clocks at both ends – this is known as synchronous transmission and nothing more will be said about it.

If serial transmission seems hard to imagine, think of Morse code. When SOS is sent as Morse code, the first S is sent as three dashes or long notes. In Morse code, any letter can be made up by combinations of dashes and dots, sent in order. A word is just a series of letters. The same principle is used for computer serial tranmission. Letters are sent, not as dots and dashes but as zero or ones. This is convenient because letters are also stored as combinations of zeros and ones inside all computers; the difference is, all the zeros and ones are sent together.

For a serial transmission the zeros and ones are sent out in sequence. Unlike Morse, where there is a silence after each dot or dash, on a serial line only zero or one can be sent. To show when the bits are starting to come a signal is needed.

In the diagram (2.1) you can see that the line was originally held in the one position. Before the first bit of signal comes a zero is sent down the line. This is just a start bit to show

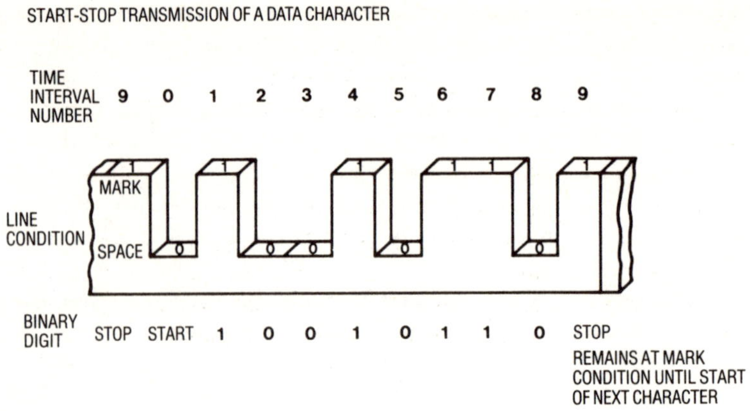

[Illustration 2.1]

something is coming. The bits are now sent one at a time; and then they are followed by another zero as a stop bit. This is the basic pattern of serial transmission.

Different numbers of start bits, data bits and stop bits can be used and these must be arranged in advance – yet another piece of standardizing.

The zeros and ones from the serial line are translated by a modem into tones which are sent down the telephone line. The receiving modem must match the sending one in standards so that they can communicate together. There are a number of such standards; these are explained in chapter six, together with the different ways in which the modem can behave. The exact way that the information is changed into tones need not worry anyone.

Everyone always forgets that each screen or computer used has its own characteristics, which means that it has its own standards as well. This isn't always obvious to the user. When a key is pressed, for example a Y, the letter Y will appear on the screen. Normally it doesn't matter what internal code is used. But if the code that the computer uses is unusual then a different make of computer might receive the right pattern of bits but show the wrong character. Screens also vary in small respects such as the width of the screen and whether lower case (small letters) can be displayed. Even such details will ruin a call if they don't work correctly. More importantly, some computers need a line feed sent after each carriage return while on others no line feed is needed. All this kind of information can be found in the maker's instruction book. While you might not need to know it, often someone you call will need to know and will ask you.

Standards include the formal agreed standards such as the speed of transmission and the kind of modem you use, as well as the actual way in which you are working – your personal standard such as how wide your screen is and whether you can receive both capitals and lower case. To be able to talk to someone else you must agree to share both types of standard.

CHAPTER 3
Making Your First Call

Imagine yourself equipped with a computer and modem, with a telephone line nearby. If you are lucky this isn't a fantasy, but you have the machinery to hand, and are keen to try it out. The steps are fairly simple and only vary a little with the machine type. You also will need to have a small program for your micro, called a terminal program. For your first few calls a simple one will do. These can be typed in from magazines, bought at a software shop or supplied free from a computer owners' club, depending on your machine type. A few simple ones are given in the back of this book for you to type in.

Making a call on your own can seem a bit mysterious. By breaking it into small stages, the more technical parts can be followed rather like a cookery recipe. If you want to find out how and why each of the steps is done skip ahead to the chapter that deals with that part in more detail. It is probably best to read through the whole process once or twice before going ahead with a call. It always seems hard to keep a place in a book while concentrating on other jobs. You will need one hand to hold the telephone and one to dial – so when you are ready prop the book up where you can see it. Don't worry about damaging someone else's computer by phoning in and then making a mess of things or being cut off. Everyone who accepts calls knows this can happen and so will take steps to make sure that nothing will go wrong or be damaged by faulty telephones or inexperienced users. Most systems you call will try to steer you back on to the right path if you make a mistake. Remember that most computers and their owners have a heavy-handed sense of humour. If they ask you to do something they usually mean it quite literally. For example, if someone asks you, where are you (Town, County)? The first step is to switch on your micro as usual and connect your modem to the phone socket, if the modem is a 'directly connected' type. The other type of modem, 'acoustic' type, which has a space for the phone handset to fit in it, doesn't need to be connected to telephone. Both kinds need either batteries or mains power.

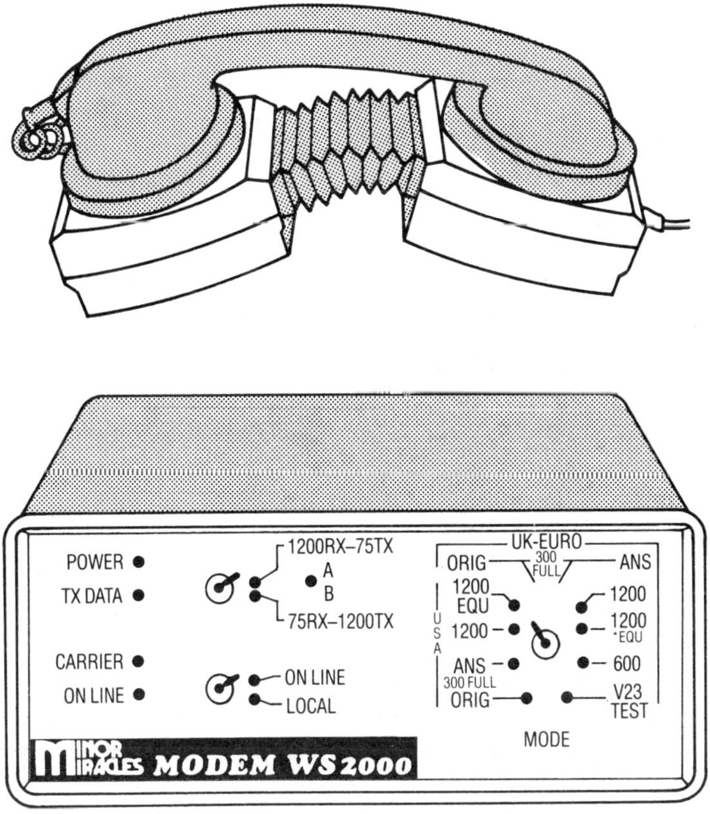

[Illustration 3.1] The Two Modem Types

Most computers have a standard connector plug to connect to the modem. This will only plug in one way round. Your handbook will show you where it is, under the name 'RS-232 Input/Output'. A few computers have unusual connectors and you might find later chapters helpful if this happens. Once you have found the socket connect the modem to your computer.

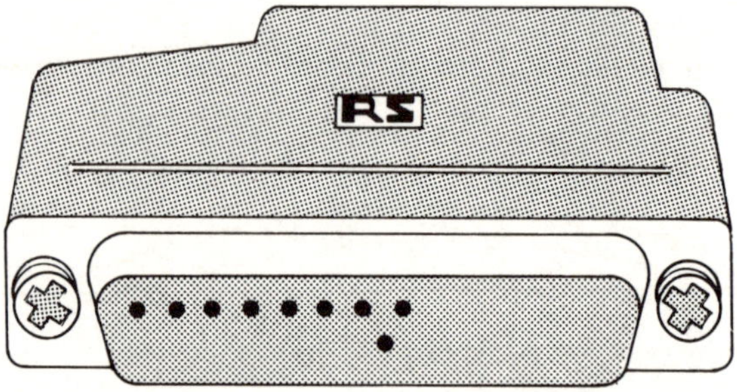

[Illustration 3.2] RS-232 'D' Plug

Load or type in your terminal program and run it. Often it will ask for a configuration or set up. Just answer the questions following as closely as you can to the example shown here. This is for a typical call to a bulletin board, and will nearly always work with any system you call. Not every terminal program will ask you all these questions, in fact, most won't ask any at all. Don't worry if this sounds like gibberish – these are the only technical bits you are likely to see for a long time. As they are usually the same they shouldn't be a problem.

'SET UP MENU'

What Speed ? 300
Full/Half Duplex ? Full
No. of Data Bits ? 7
No. of Stop Bits ? 1
Parity On or Off ? On
Parity Even/Odd ? Even

Once all is set up, you can dial the system. If you have a direct – connect modem you must make sure that any switch marked DATA/SPEECH is in the SPEECH or OFFLINE position. Hold the telephone handset in your hand as usual and dial the number you want.

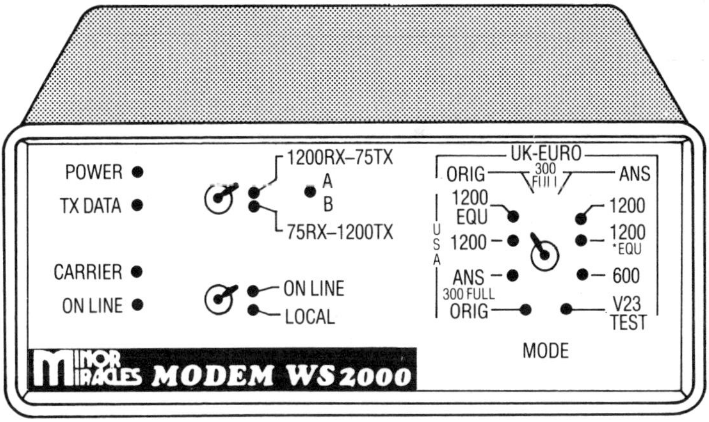

[Illustration 3.3] Making a call with switch in ONLINE or DATA position

When you get through you will hear a short ringing tone and then a high-pitched whistle. Wait for a second or two and then put the handset in the modem cup (or if you have a direct-connect modem switch over to the DATA position). If your modem has a light marked carrier it should begin to glow to show it is rccciving from the telephone.

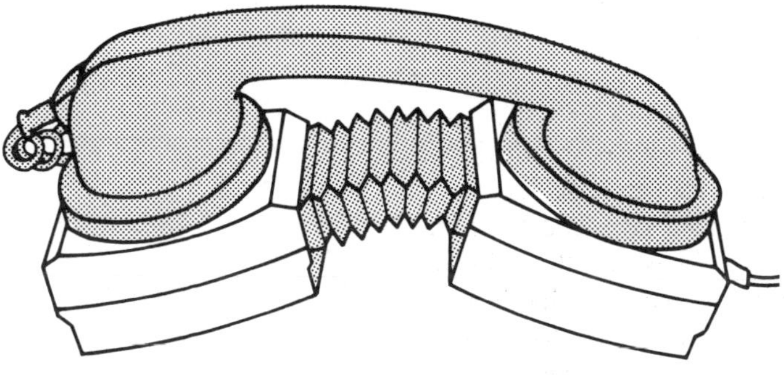

[Illustration 3.4] Acoustic Modem with Handset in Place

Congratulations. Now you can heave a huge sigh of relief. You have made your first call and connected up to another system.

What you find there depends on whom you have been calling. From this point on the steps vary widely, depending on what happens at the other end. Since there is still a common pattern about what you have to do, here is an entry method which shows the typical questions you might be asked and your likely answers. Don't be surprised if you aren't asked all these questions, or worse still if you are asked other questions. The questions will be mainly about your computer and about you. It is all intended to give you a smooth service and make sure you are remembered and recognized if you call back again. Usually you will have to answer the queries once only.

It is always possible that nothing at all has come up on your screen. *Don't panic* – this is quite correct on some systems! Type a few carriage returns. This key is called all sorts of different things on different computers NEW-LINE, ENTER, RETURN or even just CR. After a few of whatever your keyboard likes to call them, the system you are calling will wake up and realize you are there. If you still get nothing on the screen, then check that the carrier light is still glowing. If it has gone out, it means that somehow the call has been lost and you aren't connected any more. Just redial and try again. If the carrier light is still glowing and you receive nothing then hang up and start again from the beginning.

If nothing happens after trying again, you should try to ring another number. It is easy to forget that other people's computers can also have problems. Most of the systems you are calling are running unattended. Once a day the owner will come out and check them. So if there is a fault it might go unnoticed for a whole day.

Of course the trouble could be with your own equipment. Turn to Chapter thirteen – What to Do When Nothing Works, which should help you to find out what is going wrong. If you are still unable to get anything to work, your best action is to contact the owners club for your make of computer.

By now you should have found that the system you are calling has recognized you and sent a page announcing itself. There will be 10 or 15 lines giving the system's name and hours and then the screen will start to ask a few questions.

WELCOME TO TBBS No 1
OF ENGLAND
Your Local Board
***
A MEMBER OF THE
ASSOCIATION OF FREE
PUBLIC ACCESS SYSTEMS

FIRST NAME?

The first question is almost always the caller's name. Some may ask for your first name and then your last name - and others may ask for your last name and then your first. The instructions will always be clear. Follow them literally. Your name is very important. If you use a nickname or abbreviation then it must always be used on all future calls. Don't give a false name. No one else who knows you will be able to contact you if you do. The system will then repeat your name back to you and ask if it is correct. If there are any spelling mistakes or anything is wrong answer NO and press the RETURN key.

From now on, any key to be pressed will be written inside the angle signs. For example if J is to be pressed it will be written ⟨J⟩. On most systems you call, the ENTER or RETURN key will be described as ⟨CR⟩.

The host board will now ask for some information about the machine you are calling with. The questions vary, but often include asking how wide your screen is. Here they only want to know how many letters you allow on a line, not the size in inches of your picture. You may be asked whether or not Line Feeds (LFs) are needed. Some computers need a line feed after each carriage return and some do not. If you don't know, answer that LF is needed. This will stop everything that comes staying on one line, always writing over the previous line.

When all the introductory questions are over, the computer being called will show its list of choices. Here it is time to explore. Each phone number you call will arrange itself a little differently. Some special features of computers you might be calling are listed in other chapters. Type away and enjoy the call!

Sometimes it is easy to get stuck or puzzled. If this happens try typing ⟨CR⟩ or ⟨?⟩. Often this will produce a little help or a few notes. Even just typing ⟨HELP⟩ can work wonders on some systems.

Finally the time will come when you have to end your call, often all too soon. The 'Menu of Choices' on your screen will usually have one marked Goodbye, Quit, Terminate or something similar. Choose that and follow the instructions. Don't hang up yet! When the computer has signed you out it will say goodbye and often ask you to hang up your phone. The carrier light on your modem will go out and then you should hang the handset up on the telephone. If you have a direct connect modem put the VOICE/DATA switch back to VOICE or OFFLINE.

Now the call is finished, you can call yourself a Hacker. Turn off your computer and monitor, as well as the power switch on your modem.

CHAPTER 4
Software Review

Without software any computer is a dead piece of electronics. This is especially true for communications work. Everything can be connected electrically and yet there might be no way of actually linking it all together for a phone call.

The software for communication is generally called terminal software or terminal program. This is an accident of history which arises from the fact that originally a keyboard and a printer would be used to work a large computer – this combination being called a terminal. Many of the terms used still come from that time. The early programs only mimicked a keyboard and a display. That means that what was typed was sent, and what was received from the modem would be shown on the screen. This kind of software works, but is not very convenient. Some simple examples are given at the back of this book to help you when you are getting started. These programs are known as dumb terminal programs because they cannot do anything themselves.

For convenient use other features have been added to the basic dumb programs to make them into smart programs. The smart or intelligent program cannot do any thinking for itself either. The intelligence just means that it does not behave like a typewriter, but has extra features built in.

There are two main methods for writing and operating programs: menu-driven and parameter-driven. Menu-driven programs are those which show a list of choices at each option, and then ask you to select which choice you would like. Parameter-driven programs are those which run by themselves the whole time, but when you want to change a feature some special keys must be pressed, not when the program asks you to press them but when you wish. At first, menu-driven programs seem easier to use as the choices are always put forward to remind you what you can do, but after a time many people find it tedious and boring to have to wait to go through a whole menu. The parameter-driven programs are a bit harder to start with, but more convenient for regular users. Can you imagine a word processor which showed all the possible commands each time you only wanted to rub a letter out? It would seem useful at first but it would very quickly make a simple typing job slow and tedious.

A huge variety of features can be found on the smart type of programs. Many of these will be useful to some people and not to others. The more common features are explained below, but there are other special features in some particular programs – these are not explained here but they are usually emphasized in the advertising literature. When choosing software, however, it is more important to see that all the features you think you want will be there than to have a large number of frills which will never be used. Your task is to choose the most suitable program for your needs. There is no 'Best Program', just one which fits your present and future wants. Don't go to a supplier and say 'just give me the best program'.

Documentation.

This is the first thing to look for. A good program without clear instructions is a contradiction in terms. The program might be very clever, and able to everything in the hands of an expert but unless *you* can use it, it must be classed as useless. This might seem harsh but experience always bears it out.

Ability to Alter Serial Settings

This means that you can alter the word length, transmission speed and other variables in your serial output to the RS232 or RS423. These were briefly explained earlier. This feature is a must for any sensible program. Different numbers you call will have their own requirements and you must be able to fit in with them. Some types suggest you alter the program before it is run, then save the new version and run that. This is not very convenient. Most programs offer the more handy approach of allowing you to alter specifications from the keyboard. You should be able to alter the Baud rate (speed), the data word length, the number of stop bits and the type of parity being used. Some programs only allow you to switch between two preset variations – this is the absolute minimum standard acceptable.

Automatic log-on

A handy feature is to be able to preset your name and the entry details that you have to enter each time you call a board. This saves time (and so money on your telephone bill) as well as stopping mistakes. It is not essential, but where it is installed it should be practical. As well as sending any text you want, the program should be able to wait for a particular character to come in before sending the next line and also to pause for a few seconds. This allows answering questions like:
'IS THIS CORRECT (Y/N) ?'
All the software has to do is wait for the '?' and then send ⟨Y⟩.

Echoing to a Printer.

Even if you have not got a printer at the moment this feature may well become essential to you. When making a long call information is often forgotten (you may be younger and more alert than me but it will still happen). A printed record of everything that has been sent and received is very useful and can save many phone calls back to look for missed items.

Recall of previous lines.
Not common, but very useful, is the ability to have the last 100 or so lines stored in the memory of the computer all the time so you can look back at something which has gone off the screen. This is really valuable when there are lots of commands available and you have forgotten which option to take next.

Saving a buffer to permanent storage.

Just as printing is useful, so is the ability to store everything into a buffer (or storage space) in the computer's memory. This buffer can then be saved onto something more permanent such as a cassette tape or a disk. This will allow you to store text and edit it later with a word processor or even allow you to download or catch programs which are being sent to you. For this last feature the better programs will automatically open and close the buffer at the right time so that the program you receive is not mixed up with other information coming in to you. The AFPAS standards in Appendix B give the technical standards which the software should follow.

Loading a Buffer from Storage and Sending it.
This feature allows a message or a program to be written before any phone call is made, and then loaded into the memory to be sent. This means that when a phone call is made the text – your letter – can be sent at high speed, faster than you can type. Again the AFPAS standards in Appendix B will show what standards (these are called protocols) should be used.

Ability to Generate All Control Codes.
As well as the alphabet which is used for reading and writing, some special symbols have been agreed internationally as control codes (ASCII). For most connections they won't be needed but every now and then they will be necessary. This feature is essential for most users eventually.

Error Free Protocol.
We all make mistakes and the computer is no different. Noise or disturbances on the telephone line can cause letters to be lost or changed during the transmission. This does not matter too much for ordinary English text as it is easy to guess the wording from the context, but if you want to transmit figures or a computer program then just guessing isn't good enough. There are a number of methods of sending the information down the line and checking that it is received correctly. One method is called 'Modem Protocol' and was devised by Ward Christensen in the United States. The method works well and is not copyrighted,

so it is available for many different makes of computers. It is explained in more detail at the back. All the public boards in Britain have agreed to support it. Whenever you want to collect some software from a bulletin board it is the best method.

X-on/X-off.
This sounds mysterious but is really very simple. The 'X' stands for switch. By sending an agreed character you can instruct the computer at the other end to stop sending you information immediately. This might be because you want to print something and the printer is slow, because the text is moving too fast to read or for any other reason. The detectives amongst you will have reasoned out that the X-On is the instruction to restart after having stopped. This system can work both ways – it might be that you are going too fast for the host computer and it will try to stop you, to give it time to catch up.

A Variety of Uploading Methods.

As well as sending text in a continuous stream, you can stop after each line and wait for a signal for the next line to be sent. Some systems prefer this. The signal can vary – it might be an X-on, a '?' or most commonly a ⟩ sign. The better programs can cope with all these facilities.

All users have their own particular requirements but most or all of the above features will satisfy nearly everyone. There is no need to spend a lot of money on terminal software. Many good programs are available at low cost. Some of the best are available free. Some software authors have generously given away the copyright to their programs, and the software is called public domain. This means there is no charge for the program. This doesn't quite mean they are free. The copyright is not the only cost in a program, there is also the cost of distributing it, and sometimes of making small changes to fit a particular make of machine. Public domain software is not always free - but almost always you can make and distribute as many copies as you like. In choosing a program to use some features are essential for everyone, others are helpful for special applications and some are just there to make life a little easier.

In the first group, the essentials, I would include the ability to change the 'RS-232' parameters and the Baud rate or speed. The possibility of sending control codes and of accepting the X-on/X-off system is also necessary. Without these, you won't be able to call some places. Nothing is more annoying than to call and find that nothing works and all you can do is hang up and forget all about it!

The special applications features always come up some time so it is not wise to try to do without them. They include the ability to load and save a memory buffer to a cassette recorder or to a disk system, the matching commands to send and receive from the buffer to the modem, and to use the prompted line-at-a-time sending method to help slow receivers at the other end.

The rest of the features are optional. All of them are worth having and will make for simple quick calls, but you can do without them. For example, there is no need to have an automatic log-on; it is just nice and easy to have it.

Some of the programs in use are very common and you are likely to meet other people using them. It helps to know what they are like so that you understand what is happening.

MODEM7.COM is probably the most common public domain terminal program. It is written for microcomputers using the CP/M operating system. It is full of very clever features which are rarely used. It often has to be configured or adjusted for each make of machine. To do this a good programmer is needed who understands machine code. When the program is installed it is able to act as a simple terminal, save text to a buffer and then to a disk file, to transmit and receive blocks of text written at a different time, to transmit and receive files using the modem protocol. It is a good program which is a little hard to use at first. It includes the support for an automatic telephone dialler if your equipment supports it (this is pretty unlikely).

STERM/CMD is a public domain program written for the TRS-80 and The Video Genie range of microcomputers and is excellent in every way except that it does not support the modem protocols. It will work out a checksum for a file being sent or received, but this is often not much help. Like modem7.Com it is a command-driven program. It is unusually good when copying text to a printer, as it will hold some text in a buffer and send it slowly to a printer, without slowing down incoming text.

JTERM is the equivalent public program for an Atari computer. It is a little less advanced than the two mentioned before, but it does accept the special Atasci coding method which is sometimes used on Atari computers. It does not seem very widely used, and offers very little control over the Baud rate and RS-232 parameters. It doesn't support the modem protocols nor can it calculate a check sum. If this sounds off-putting don't be discouraged, most users find it excellent for everyday (or every night) calls.

COMM-STAR is a widely available program for the BBC Microcomputer. It is available as a ROM chip which only has to be plugged in, and is easy to use. The author has implemented the modem protocols very cleverly to send data with other computers, and also to transmit BBC machine code programs together with their address in memory and starting memory location. It is a popular program that has been written with the amateur in mind. There are other ROM based programs available but they are really trying to imitate particular types of display terminal. In everyday use they are not as convenient for ordinary hacking and communications.

Some very simple programs are given in Appendix A. They are only meant as quick short programs to start you up. They are missing many of the better features that by now you will expect and hope to have.

CHAPTER 5

The Equipment You Need

Communicating with the outside world involves a chain of equipment starting with your fingers and ending with the receiver. Luckily, we only have to worry about going as far as the nearest telephone.

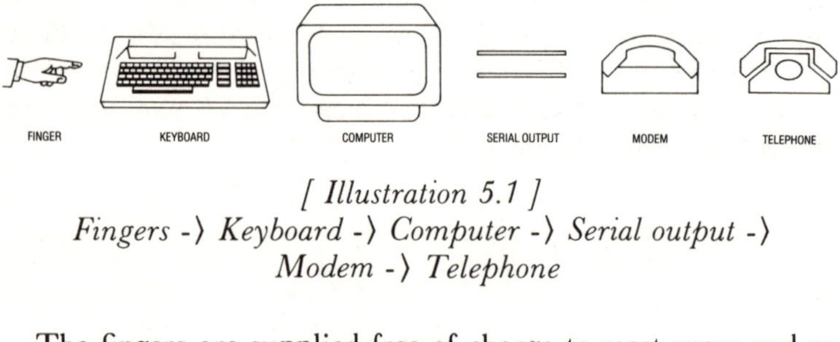

[Illustration 5.1]
Fingers -> Keyboard -> Computer -> Serial output -> Modem -> Telephone

The fingers are supplied free of charge to most users and so won't be discussed further. All the other parts of the chain must be present or a substitute found, not forgetting the connecting wires and the plugs. It is easiest to examine one stage at a time briefly. Detailed information will be found elsewhere in the technical chapters to explain exactly what each of the stages does.

You probably already have a microcomputer if you are reading this book, but a computer is not essential and other pieces of equipment can be substituted for one. If you are one of the few who have larger computers then parts of this chapter are even more necessary. Just because a computer is large or powerful does not mean that it has the features which are normal on smaller, more specialized microcomputers.

Your computer will have (at the very minimum) a keyboard, a display and a central processor and memory unit. For most programs these are the only parts that you will see. To get your computer to be able to talk to the outside world it must be connected up. Look in the maker's instruction book to see whether it has a serial output. This might also be called RS-232, RS-423, serial port or even communications line. If your computer doesn't have one of these, don't despair – the manufacturer may make one as an external box which can be attached – but generally this is only done with very cheap computers.

By now you will have found the connector for the serial line. Your phone will probably also have a plug which goes into a wall socket. In Britain there are two types which you might have. The older type of telephone plug is round and looks like a giant model of a ballpoint pen. This fits into a wall socket which just looks like a round hole. The newer-type plug looks like a small square sugar cube with silver pins on one side. It fits into a neat small box which has a little shutter over the hole where the plug goes. Unfortunately you can't connect the the computer direct to the telephone system but you have to use an intermediate box called a modem.

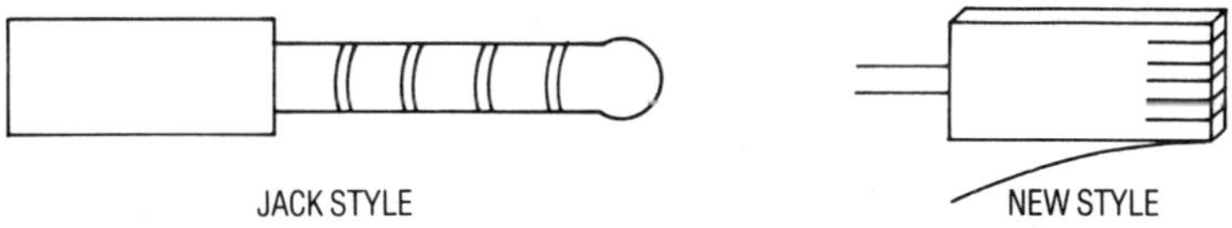

[Illustration 5.2] Telephone Wall Jack Sockets in U.K. Use.

The modem is the essential part of the system. You can do without a computer or a display screen but you can't do without a modem. Modems turn the computer information into sounds which can be sent down the telephone line. Most modems are wired into the phone systems with one of the two plug types described. These are known as direct-connect modems. Other modems have little loudspeakers and microphones inside them.

They allow the telephone handset to be fitted snugly, and send the sound down the handset or listen to the sounds coming from the telephone handset microphone. These modems are known as 'acoustic-coupled' modems. If you haven't got one of the telephone plugs described then you will have to use an acoustic modem. They are usually not quite as good as the direct-coupled type but they are easier to instal and more portable. A warning – acoustic modems won't normally fit the new-style electronic telephones as the handsets are designed differently.

The modem is connected to the computer serial line through a wire with plugs at either end. Normally there are few problems if the computer supplier sells cables, but there are some hazards to look out for. The connector at the end might be a socket rather than a plug. The modems almost always have sockets built in, so that the cable needs the male plug to match. Some shops sell special adaptors which will make a plug into a socket – they are called gender reversers. The other problem is that the wires in the cable may need to be swopped round. Usually this can also be fixed by fitting an adapter called a 'null modem'.

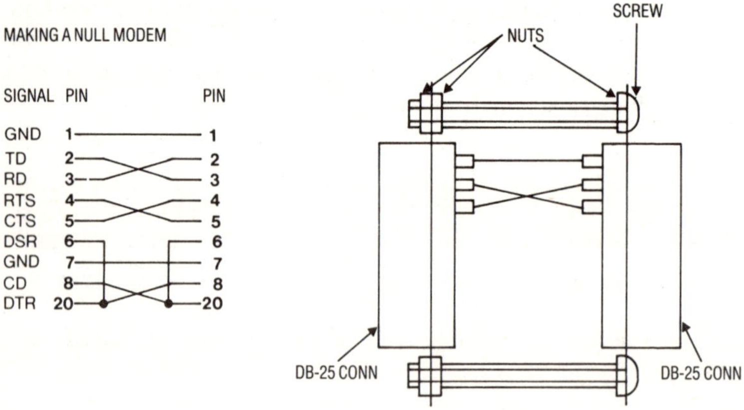

[Illustration 5.3]
A Null Modem and a Gender Reverser

Your computer will also need a special program called a 'terminal program' to make it usable for communications. There is a wide variety of these available and the selection can be quite difficult. It is probably easiest to start with one of the very simple terminal programs which are available free of charge.

The computer-modem combination is the most common method of calling the world, but it isn't the only method. A few computers are now available which have a terminal program permanently loaded and some even have built-in modems. This is increasingly common as it has given rise to small portable sets which can be used anywhere, even in a telephone box!

The terminal program turns a microcomputer into a terminal, an old term which covers mechanical teletypes as well as modern VDUs (Visual Display Units). Either of these devices can be used without any problems instead of a microcomputer. In fact, most commercial users would use a VDU terminal with a modem rather than a computer. A teletype is a type of mechanical typewriter complete with keyboard which also has electrical input and output. Generally they are not very good for communications, as they are rather slow and noisy. A special word should be put in for the little silent terminals with built in printers and modems. These fold up to make a small attaché case and can be carried around and used anywhere.

There is no need to worry about compatibility of different makes of equipment – the modem need not be from the same maker as the computer. As long as your computer has a serial output labelled RS-232 or RS-423 you don't have to worry. The standard says what voltages are allowed and how the connections work. Quite a lot of secondhand equipment is available at very low prices after being sold by British Telecom or commercial users. As long as the equipment has an RS-232 input it can be worked with a matching microcomputer.

There are, however, two incompatible types of modem used in Britain: those working at 300 Baud (about 30 letters per second) and those used for Prestel which receive at 1,200 Baud (about 120 letters per second) but which only transmit at 75 Baud. The Prestel type is generally of less use but more widely available than those at 300 Baud. Don't buy modems in the United States for use in Europe as they are designed to work at different frequencies and can have other complications.

CHAPTER 6
Modem Standards

So far the word 'modem' has cropped up in nearly every paragraph, which all goes to show how essential it is. Modems come in a wide variety of designs and capabilities. Every month new ones appear with new features to appeal to buyers. Most of the features are variations on the common options available.

To recap, the modem or MODulator/DEModulator takes computer data coming in and turns it into tones which can be sent down the telephone line. It will also do the reverse - receive tones and turn them back into computer data. The modems at each end of the telephone line have to match in standards for each to be able to receive the other's transmissions.

Modems can be divided into two main classes - directly connected and acoustically coupled. The directly connected class consists of those modems that are wired into the phone system and send their information through those wires. The acoustic group consists of those which have no electrical connection to the phone system.

Generally, direct-connect modems are preferred as they are more reliable. Without microphones and loudspeakers, they are less sensitive to local noise and vibration in the room where they are. Most require a mains electric power socket to run; a few run off batteries and some unusual types use electric power taken from the telephone line. The direct-connect modems often have many extra features which are missing from the acoustic type.

Acoustically coupled modems rely on actually picking up the sound from the loudspeaker in the handset and transmitting sound to the microphone. Usually the handset has to fit well in small rubber connectors to make sure that the sound is efficiently transferred. Acoustic modems don't often work with the non-standard and fancy phones. Because they work by picking up sound, acoustic modems can be less reliable if there is noise in the room around them but, for all their disadvantages, they are still very popular because of their portability and easy installation.

Modems work by turning the serial computer data into sounds. In this book only asynchronous modems are going to be examined. This is the class of modems generally used for amateur communications. If you are offered a synchronous modem (the other type) you will probably not find any use for it.

The modem can send one of two tones – let us call them high and low tones. When the computer sends a 0 signal, the modem sends a short burst of the high tone. When the computer sends a 1 the modem sends a low tone. The receiving modem decodes the tones back to zeroes and ones. When there is no

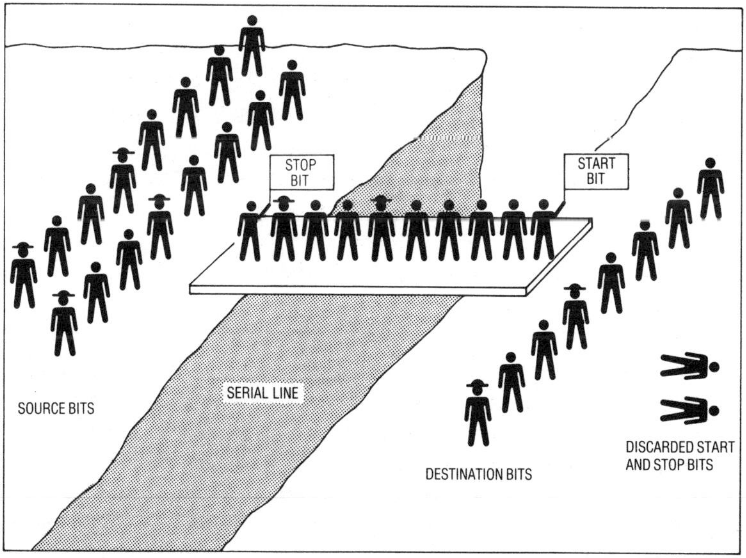

[Illustration 6.1]
Diagram of a letter A being sent.

information the tone stays on in the 1 position. If the transmission was done this way it would be hard to know when a signal which was all ones was sent. It is normal with the modems we are looking at (asynchronous or without clocks) to have a start signal to say something is coming and a stop signal to say that the few signals have all been sent.

The most common practice is to follow the international standard V.21 . This allows two frequencies (the high and the low) to be sent from person A to person B. Person B has also got two frequencies to transmit to Person A. All four frequencies are different; this means that A can be sending to B while B is sending to A. The people A and B are usually called the 'originate' and the 'answer' positions. The originate end has two frequencies which he transmits on – the originate frequencies – and two frequencies which he listens on – the answer frequencies. The answer end has the reverse case – transmitting on the answer frequencies and receiving on the originate frequencies.

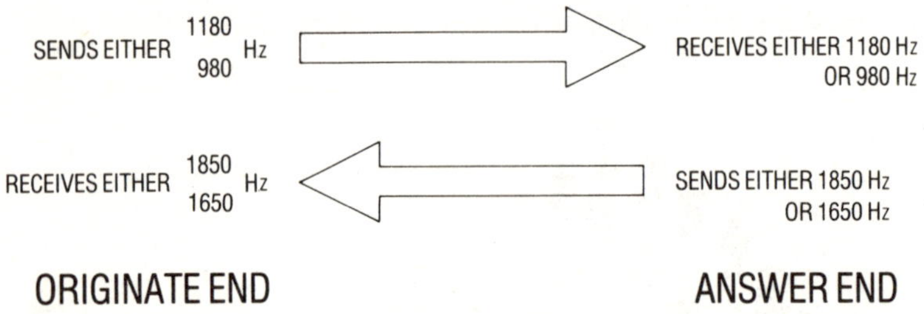

[Illustration 6.2]
Forward and backward Channels.

A modem which is called 'Originate Only' can only send on the originate frequencies and receive on the answer frequencies. Modems called 'Originate/Answer' can work either way round. It should be understood that an originate modem can only communicate with an answer modem, not with another originate modem. This means that if you and your friends have originate-only modems you cannot directly communicate together. Originate modems are generally used by people who call into large systems.

To complicate the matter still further there are two incompatible sets of frequencies used throughout the world. The United States, Middle East and Japan use tones known as Bell Standard. All of Europe and the rest of the world uses tones following the CCITT standard. For use in Britain CCITT standard tones are essential. A few modems coming on the market have both. Don't get a Bell standard-only modem unless you can afford very big phone bills.

The speed at which the modem will run is measured in units called Baud. This unit is named after E. Baudot, an early telegraph pioneer. Some people have thought that BAUD stood for Binary Asynchronous Units of Data. This is a wrong derivation but does explain what a Baud rate is. The Baud rate is the number of bits of information sent per second. The count includes the extra non-data bits sent such as parity bits. Most systems use 10 bits of data to send one alphabetic character. This means that 300 Baud (the commonest speed) is about 30 letters per second. This is about the fastest speed that you can read at and so is therefore quite popular and practical.

The 300 Baud speed can be with either the Bell standard, in which case it is called Bell 103 or it can be to the CCITT standard in which it is called V.21. Suppliers of modems will always state which of these two standards their modems work to.

Faster Baud rates are possible and can save time on telephone calls. For example, Prestel and many other services send data to you at 1200 Baud (about 120 characters per second). This is much faster and makes pages of text seem to appear on the screen almost instantly. Remember, however, that it will still take the same time to read the text so not as much time will necessarily be saved. The Prestel system receives information you type in at a much slower speed – 75 Baud. This is only seven to eight letters per second and can easily be beaten by most typists. This standard is called V.23 .

There are still faster Baud rates and a proper 1200 Baud, with that speed in both directions, is popular in the United States. Unfortunately, there are a number of different methods in use for this speed which are all incompatible. Many of the American modems are therefore designed to run at all the possible standards. This has meant that they have tended to be rather expensive.

Modems of this type are often called triple Modems, and can accept Bell 212, Racal-Vadic and Bell 103 calls. Here in Britain 1200 Baud transmission in both directions is not yet common for amateur use. This will probably come in the next few years and will follow the CCITT V.22 standard.

The terms full duplex and half duplex may often be used in connection with communications services. These are not entirely a function of the modem being used. Full duplex means that information is sent in both directions at the same time independently. Half duplex means that information is only sent in one direction at a time. The Prestel (V.21) is normally thought of as a half duplex system since the 75 Baud channel is very slow and wasn't originally meant for use as a reverse direction channel for data. Full duplex systems often echo your own characters back to you onto your screen. When you type a letter it doesn't go straight to your display but goes to the distant computer which then re-sends it back to you to be shown on your display. In a half duplex system, the information can only be sent in one direction at at a time, so your system will display its own characters directly and also send them to the distant machine.

If a modem is mainly used for receiving calls, it is often very useful for it to be able to answer the phone by itself. This type of modem is known as an Auto-Answer modem. If you call into a computer system through the telephone network it is almost certain that this type of modem will be used so that there is no need for someone to be waiting to supervise and answer each call; after all, computers are meant to be able to automate and do simple jobs for us! The modem is able to detect the phone ringing and take action.

Two different methods seem to be in use for auto-answer modems. Most American modems when switched on will listen to the telephone line waiting for a ring to come. When a ring is detected the modem will answer the call (Go on-line). It will then listen for a modem at the other end. If a modem is found then it will signal to the computer that all is ready (Carrier detect) and allow communication to start. If no modem is found after a preset time then the modem will hang-up (Go off-line) and resume waiting. The older method, which is more common in Britain is for the modem to detect the incoming ring, and tell the computer that there is a call coming. The computer must then signal the modem to answer the call. The modem will then signal if it finds a modem at the other end or not and it is then the responsibility of the computer to decide when to hang up.

CHAPTER 7

What There is to Call

After assembling and connecting all the necessary equipment, it's time to think about what you can call. There are three main groups of service which can be easily reached by most people through the ordinary telephone system. The easiest to gain access to are the free services provided by some companies, these are mainly based around London and the southeast. Also without difficulty are the large number of free public access bulletin boards which exist all round the country. Widely available, but requiring some advance planning, are various commercial services, such as Prestel.

As communications equipment has become lower in cost, several private individuals have sent up computer bulletin boards around the country which are available for anyone to phone in and use. These boards are free and are run as hobbies by their operators. The system operators (who call themselves 'sysops') have all grouped together to form AFPAS, the Association of Free Public Access Systems.

AFPAS boards all try to keep to similar standards to make everything easier for callers and nearly always have some features in common. However each AFPAS board is independent and will have its own features and individual areas. A message area is nearly always available for use by any callers, together with games, bulletins and news features. Users who are prepared to leave their name and address generally will find more advanced features that they can use. Private messages which can't be seen by other people, the ability to download (or receive) computer programs from a library and similar privileges are common.

Some of the AFPAS systems are completely written and programmed by the system operators, but most rely on using commercial software as a backbone for the system. This means that there are a number of systems which share common ways of working. This can vary from which keys are used to finish your call to the ways there are to enter a message. Normally the boards are named after the software. The most common sytems are TBBS systems , Forum-80 systems and CBBS systems. The fact that several bulletin boards use the same system, such as TBBS does not mean they will be the same. The areas and features provided may, and probably will be, quite different.

The AFPAS boards are usually run by private individuals purely for fun. Because of this, they tend to grow in the directions of their operators' interests. Most operators seem to be ex-players of adventure games and lovers of the type of fantasy series you find on television or radio. You will often see individual opinions and comments in reply. It is this personal touch which lends attraction to privately run systems.

As these boards are privately run and you are calling into the system operator's own home, a few hints on good manners might not be out of place here. The system operators all enjoy callers and wouldn't set up the systems if they didn't want you to call them. *Don't* be shy! If the Sysop breaks in to ask a question or to help you don't hang up immediately. If you find you have made a mistake, don't worry, usually there is a way of cancelling it – in the worst case leave a message addressed to Sysop and he will fix it for you.

Most Sysops don't like users to use false names or nicknames. This is not out of perverseness but from experience. Nearly all the bulletin boards work by using the caller's name for reference. Since most people seem to forget their nicknames or their assumed names, the result is that they don't get mail addressed to them and that many extra entries are created, wasting the system's resources. Some of the Sysops can be quite upset by this kind of thoughtless behaviour, so do your part and use your normal name. When asked for your name it is best to follow the instructions given very literally. If asked for your first name then use your first name, not your initials or your name and initials. The operators spend a great deal of time planning out features on their systems and often these need your help to work properly.

The hardest part of the sysop's life is the housekeeping and clearing out of old messages. You can do your part to help in this by deleting messages addressed to you when you have read them. More importantly if you address messages to ALL then you should remove the message when you finally get the answers that you wanted.

Obviously, you should try not offend others. Generally try to avoid personal comments and opinions on religion and politics. Most operators also take a very dim view of expletives and bad language. The boards are read by everyone, and if it is thought that you are likely to offend other users you may find yourself banned from a board. Don't think that these rules are strict – they aren't – but public spaces do seem to tempt some graffiti writers and the same thing happens with bulletin boards.

Do leave messages and contribute to the board. Remember that the sysops only provide a service; it is the callers who write all the messages and make the system interesting to call and browse through. You don't have to be an expert, but do take part and enjoy the fun.

Since AFPAS systems are run from private homes using ordinary telephone lines, the same number is often shared for the personal use of the sysop and his or her family. If this is the case you will be asked to use a ringback method. To use a system like this, you just call the number of the bulletin board and let the telephone ring for two rings only and then hang up. Straight after hanging up, pick up the phone again and redial. This time the computer will answer you. Your first two rings are used as a signal that a computer call is coming through and the computer will answer it. This stops all the sysop's personal friends being greeted with a loud tone from a modem whenever they phone to speak to someone. Other systems will answer on the first call without the ringback arrangements, but the computer may only be on for a shorter time each day.

If, when you phone in you are answered with a voice don't just hang up. Often the sysop is trying to tell you something, such as that there has been a change of telephone number or that the system is temporarily off the air. It can be very frustrating as a sysop to receive a series of phone calls where the callers hang up every time you answer the phone.

Not all the free systems are run by amateur system operators. Many are run by commercial companies, either for publicity or to sell their products and services. These boards naturally are less oriented towards the user, but have a large area for their main purpose. However, as they are often run on larger systems with many telephone lines, it easier for callers to find a free line and get into the system at busy times. The best-known two commercial free systems are Rewtel run by Radio and Electronics World, and Distel from Display Electronics. Both these systems run every day and all day. Distel is one of the first British public systems available. It doesn't have any messaging facilities, but allows you to browse through an enormous stock list of new and surplus computer equipment.

Rewtel only allows unregistered users 10 minutes on each call. There is a thriving message board and a useful forum for viewing and ordering products. If you register and pay a small fee you are given a longer time on the system and there is more that you can do.

The most widely known system in Britain is Prestel. This was originally part of British Telecom, but is now a separate company. It offers a nationwide system for calling and retrieving information. Prestel has limited message facilities but these are being expanded all the time. Normally Prestel runs at the 1200/75 Baud standard, which is different to most amateur boards, but there are facilities in London which allow users to call in with the more common 300 Baud standard modems. Also when the 300 Baud Prestel service is used, the graphics characters are removed and replaced with stars so that almost any microcomputer or terminal can be used. If you are interested in this special 300 Baud service it is best to contact Prestel headquarters in London who will make all the arrangements for you. There is no extra charge for this service.

The way forward is probably best illustrated by the large networks available in the United States. The two best-known examples of these are The Source and Compuserve, both very large bulletin board systems and information storage systems similar to the reference section of a library. Much of the information is in specialist areas, although it might take a little time to get used to wandering through all the corridors and shelves so that you don't get lost. Both these systems run on very powerful mainframe systems and handle hundreds of callers at the same time.

As these systems can handle many callers simultaneously it is possible to have conversations with users. This can be so like amateur radio or citizens band radio that Compuserve actually calls this spot the CB Area.

The companies that operate these systems have installed their own telephone lines from many cities in the United States back to their main computers. This allows many users to be able to call in at low cost local rates. Perhaps this facility will come to Britain soon.

It is worth remembering that the traditional Telex system can also be reached with a microcomputer and a modem. For occasional users, one of the large London accommodation address agencies offers a service where you can use your computer and modem to phone into their computer, which will then automatically send on your telex. and also store your replies to await your collection. This has the extra advantage that they will telephone you if they receive a telex, and then read it to you. Other routes to send and receive telexes are through Prestel, through the Packet Switch Stream service, or through British Telecom's Gold Service.

CHAPTER 8

More Professional Areas to Call

All the systems described in the last chapter were oriented towards the private user calling from home. Their charges were low, or free. Some were run as publicity aids, others as hobbies. This is not the normal way for any enterprise to work. There are a large number of major systems which are run commercially or by educational establishments such as universities, polytechnics or libraries. These systems are not designed for the private caller, but they will all make their services available either for a small charge or when they feel that you are an experienced and serious user who will not damage or interfere with their normal workload.

One of the most widely available systems in Britain is British Telecom Gold. This is a large bulletin board-plus service available throughout most of the country, designed for use by businesses and offices. Normally each business forms a group and members within the group can use the service to exchange messages, or even talk directly with each other. The system is very advanced and has many clever features. Messages can be sent express so that the addressee is told to see them urgently and calls may be placed through BT Gold to British Telecom's pocket radio-pagers, to tell a user that there is a message waiting for him or her. The system will even check the spelling of the messages before they are sent.

Just in case these features aren't enough, BT Gold will do

many other jobs. It can be used as a word processor. Files containing information may be stored on it (for which there are, of course, storage charges. The system may be used as an appointments diary, and will remind you of appointments and meetings when you call. The main software for BT Gold was bought in from suppliers in the United States, where it runs under the name of Dialcom. There is good access in all directions between Dialcom users in the United States and BT Gold users in Britain – they can easily send messages to each other and the system will look after all the problems of routing and dialling. BT Gold has good connections of this type with most of the world, and is often a good choice for business use. For private users, there is less to attract you. The system has a number of games and other amusing features, but similar and better are available on most free public systems.

In the United States two large networks have grown up which are commercially run and which do encourage the private caller as well as the business user: Compuserve and The Source. In many ways the two systems are similar. They both run on large powerful mainframe computers in one central location. They have however linked themselves to the telephone system to allow local calls to be made from most American cities to contact their mainframe computers.

Compuserve follows the pattern to be found on most bulletin boards, but it also offers a unique CB area. This allows a caller to chat with some of the hundreds of other callers who may be using the system at the same time. The system's convenience makes it a real pleasure to use. You can use the CB area with a pen-name if you want to stay anonymous and you can also set it so that you may speak to anyone *except* certain people who you may want to avoid. For privacy messages may be sent scrambled, so that other people who can see what is going on can't read the message text.

If that was all Compuserve offered then it would not be a great advance, but it also supplies an enormous library in special interest areas. The most appealing part of the Compuserve system are games such as Dec Wars. Unlike most computer games these don't pit your wits against a machine or a program. Your opponents or perhaps assistants are other players who are logged in at the same time.

Unfortunately (or perhaps for those paying the phone bills, fortunately) Compuserve is not normally available to those calling from Britain. If you call direct long distance to the United States, the modem frequencies are different and you will not be able to access the system. Compuserve is, however, available through the International Packet SwitchStream network (IPSS). This is a relatively economic way of calling the United States and access charges average about five percent of direct telephone dial charges.

The Source is a similar installation, owned by Reader's Digest. It is slightly less trivial in what it offers (or, others might say, less fun) but has an excellent range of specialist libraries and encyclopaedias. The Source has no access points in Europe and is usually reached by calling (expensively) long distance or by using the same International Packet SwitchStream Service. However, The Source has fewer complications when it comes to registration if you are not a resident of the United States.

Multi-user games are available all over the world, not just in the United States. There is an excellent multi-user Dungeon game called Mud (also one titled Valley) which has been running at Essex University for some time. The game runs in the evenings and weekends when the system is not in full use and has many players from all over the world playing it. It can be accessed either through the IPSS network or by direct telephone dialling. As the system is limited in facilities and non-commercial, there are strict limits on the number of players who are allowed access. If you are a regular user of the AFPAS bulletin boards then you will occasionally find details of how to apply when vacancies are available. A commercial version of MUD is expected from Century Communications Ltd some time in 1985.

Prestel is one of the most widely available of the professional systems and was briefly described earlier. It is a giant data-base consisting of pages of information of graphics or text. The organization of Prestel is unusual in that it was originally devised for users who had no full keyboards with letters but only had numbers. It is designed as a series of numbered pages, arranged in a tree-like structure. This means that you travel down a main trunk and are then offered choices of different branches. Further branches may be offered all the time as you travel down the system. This system can make it slow and tedious to find the information you want. If you don't use a printed index, then finding a simple bit of information such as the time of a train can be a slow process where you need 10 or 11 separate pages before you find the information. However, the information is there and is generally free, making Prestel quite useful overall.

Within Prestel there are some closed user groups: private areas which require authority to enter and read. Some are for commercial users, such as travel agents, others are for general users. Micronet 800 is the best known of the general closed user groups. It offers magazines, computer software that may be downloaded, and a more interesting set of pages less oriented to official information or advertising. Joining Micronet 800 normally includes your fees towards joining PRESTEL.

As well as these well-known electronic publishers, there are many others which might be of interest to some people. Examples include the British Library which maintains an on-line library search computer service called Blaise. Reuters has a system which offers press news or financial information. British Leyland has a company called Comet which provides a very professional message handling service, although it is not likely to be available, or of interest to, the amateur user.

CHAPTER 9
AFPAS Boards

Most of your calls will be made to AFPAS boards. These boards are all privately run by individual operators and tend to take on the flavour and choices of the operator. Because the boards are usually run from private homes the numbers frequently change, and the systems are sometimes out of order as they nearly all run unattended. In spite of these drawbacks they have become very popular.

TBBS London, one of these boards, sometimes receives over 100 calls per day. There are over 20 hours of calls in each 24 hour day. These systems are popular, it may be difficult to get through. The only advice is to keep on trying, preferably outside the popular hours during the afternoons between two and five o'clock and in the evening from about seven to midnight.

The AFPAS boards are all designed by the operators. Each board is different in detail, although the groups of boards using the same software will have an overall similarity in how they work. The first privately run board in Britain was Forum-80 in Hull. This system runs on a TRS-80 (as do most other boards). The operator is called Fred Brown and all modem users in Britain owe him a large debt of gratitude for starting such facilities and doing so much to encourage other people to take on the same unpaid work. Forum-80 owes its name to the series of programs which are used. The system is very advanced and has many clever features.

When you first call in, the system will ask for your name and location and then check in its records to see if it knows you from a previous call. It will ask you various questions, such as your password if you have called before. After using it a few times note down exactly what your usual answers are. Forum-80 will allow you to send a whole string of commands or answers into the system so long as they are separated by semi-colons. Forum-80 may upset new users because it presents menus and sub-menus without a full text, only the initial letters of choices are given. However, pressing ENTER or ⟨?⟩ will give a fuller list of choices and repeating the question mark will expand the choices and give some help.

Like many of the AFPAS systems part of the board is open to all, but other parts require you to register. This may be done by leaving a note on the board, and answering various questions. There is no need to worry about giving your name and address. None of the AFPAS boards will disclose your name and address to anyone. However, other registered users can see your name and home town.

Forum-80 is particularly interesting in its message entry. Firstly, users should be warned that it cannot accept messages which have been prepared beforehand and then sent at high speed. If you have such a message it is best to send it with some extra-nulls if your terminal software will support it. Messages may be sent as private or public. Private messages may only be read by the sender or the receiver, whereas public messages may be read by all. Messages may also be sent so that they are automatically removed from the system after a certain date. If in doubt use this option and the option for messages to be automatically killed on receipt. They help the sysop considerably by making it easier for him to do his housekeeping and remove obsolete messages.

Forum-80 is unusual in that it operates to both CCITT standards at 300 Baud during the day and evening, but switches to the American Bell standard after midnight. The same message base is there for both sets of users. This allows contact and messages to be sent between European and United States users, something that cannot normally be done because of the incompatibility of the two different modem modulation methods.

TBBS London was the first board in Europe to use the

TBBS software. This system is unusual in that it allows the board to be designed to the owner's wishes completely – everything about it can be changed. Luckily for callers, all the TBBS boards have chosen to keep roughly in step. Being in the London area where there are many users, TBBS London is one of the most heavily used boards in the country. New users are allowed only a limited time on the system and partial access to it. After completing a registration questionnaire users are given more time on the system, and more access.

TBBS London is divided into areas which are each accessed by their own menus. Unlike Forum-80, a full menu is normally displayed, along with an explanation of what each choice consists of. However, sending and reading all the text is slow and boring and users can change the amount of text and help shown to them, and the system will remember what they like for their next call. TBBS London has areas for various common microcomputer types and normally you only see the areas which interest you to save time.

TBBS London seems to have developed a speciality towards communications and modem problems and has several areas devoted to this. Other areas which seem popular are the sections giving details of trade gossip from the microcomputer industry and the magazine section which contains non-computer oriented articles written by outside contributors. Several adventure games are available to play on the board, one of them being very large indeed, with more 'rooms and locations' than can fit into the memory of even most large microcomputers.

TBBS Liverpool is organized and run by Peter Tootill. This board also runs on a Tandy and was the first in Britain to offer a 24-hour service. Like all TBBS systems it has its own menu design and board layout. The operator of the board acts as the coordinator of AFPAS board telephone numbers for Britain and usually maintains the most up-to-date list of where you may call. The system has an unusually useful library of programs which may be collected or downloaded by calling in with your computer. No charge is made for these programs. The board offers a full messaging service and has a special facility for sending letters and queries to one of the national monthly computer magazines.

TBBS Blandford is a Dorset-based board using a TRS-80

which also runs 24-hours a day. The board is sponsored, but not run, by a Blandford-based computer company. The system operator has a strong sense of humour and several chain letters can be read and joined. They are all humorous and based round original subjects which have long since been forgotten. Being sponsored by a microcomputer company means that the company has a section where orders may be placed, and questions asked. This is a novel idea, and the low overhead costs involved are passed on in some very low prices.

The board is operated from a site which was once an old mushroom farm, and this legacy shows itself in the board. There are cryptic references to cellars, funny farms and fertilizer which may puzzle some newcomers. TBBS Blandford is not really run by a part-time scarecrow from the back of an old tractor, but it does keep a strong country flavour and a whiff of the farmyard is sometimes in evidence.

CBBS North East is a bulletin board run jointly by two system operators in Tyne-and-Wear. Unlike the systems described previously, it runs on a Superbrain computer. Like Forum-80 it runs on European standards in the mornings and the evenings and on American tones during the late night and early morning. At first use, CBBS North East may seem to be a difficult board to manage. The style of the commands and prompts given are rather short and terse but this is just what makes the system quick and easy for those familiar with it.

CBBS North East has some interesting special features, such as the ability to search through messages looking for a keyword. The system acts as a center for the distribution of the CP/M user group software and for answering technical problems about the CP/M operating system. For security the owners give little access on your first call. If you follow their procedure, registration only takes a day or so and is free of charge.

CBBS South West is run from Exeter in Devon. It operates on a Nascom computer and specializes in CP/M software. This board is able to accept calls from either V.21 (300 Baud) or V.23 (1200/75 Baud) modems. The operator, Boyd Hitchcock, is often around to give help and advice to newcomers.

There are other CBBS systems in Surrey, the Chilterns and Cumbria, all running the same software. The boards are not identical but are very similar to use.

The Southern Bulletin Board is unusual in being run on an Atari computer. This is a 24-hour board giving a service to all computers, not just for Ataris. The software kept for downloading is mainly Atari-biased.

CHAPTER 10
AFPAS and UK Standards

In 1983 all the operators of privately owned free public bulletin boards met to plan and agree how bulletin boards could be encouraged. All of the system operators found they had similar interests and similar problems. Everyone agreed that the first thing that should be done was to agree on common features and standards so that callers would not have to change everything each time they called a different board. All the operators agreed to join to form the Association for Free Public Access Systems (AFPAS).

There was no requirement for the boards to be privately owned or run, but the emphasis was that the boards should be free and available for anyone who wanted to use them. All British boards belong to AFPAS.

AFPAS serves three main functions. Firstly, it ensures that technical features don't vary too much from board to board. In the United States and some other countries different boards had followed different technical standards and it had become difficult to call several boards. This was specially true when it came to transferring programs from a system to a user's computer.

Secondly, AFPAS has become a public relations outlet allowing the bulletin board movement to publicize and encourage the use of these boards generally. This is paricularly important for first-time callers who would otherwise never find out about the existence of free public systems at all.

Finally AFPAS has allowed bulletin board operators to approach national authorities to make various facilities available and ease the problems for telephone users. As individuals the system operators have no influence, as AFPAS they represent the interests of over 10,000 individual users.

AFPAS is a very loose organization and usually meets twice yearly to compare notes and exchange solutions. All the system operators have got to know each other through these meetings and this is as pleasant for the system operators as calling a bulletin board is for the users! It can be a great relief after struggling with problems to meet and find that others have the same troubles. The relief is increased even more when a solution can be achieved together.

The question of the need for standards seems to arise everywhere. To a user it can seem very straightforward to call in and use a board but there are many variations at each stage, and the fact that it seems common and easy to do things one way doesn't mean that it happened by accident. Every aspect has to be planned and agreed. To give a simple example, consider the size of the serial word used. AFPAS suggests that generally an eight-bit word should be used with no parity bit and one stop bit. This was chosen because it would allow the exchange of files and programs with a system which would allow automatic error detection. At the same time, there are a few computers and terminals from professional installations which have to run with a seven-bit word and even parity. The AFPAS standard would still allow these terminals to be able to use the systems. Small considerations like this have to be considered when setting up a system. If you, as a user, don't have any trouble it is a tribute to such good planning.

The AFPAS standards are given in Appendix B. They fall into three main groups: details of the serial transmission method, details of the file transfer methods and details of special function characters which are reserved. The standards were not designed by AFPAS but selected by choosing the best common practices in use.

The serial transmission standards are essential. If you are using the wrong modem type or you are set to the wrong serial details then you will not be able to make any contact at all. The standards say what frequencies, speeds and switch setting should be expected.

The file transfer method standards are becoming more important. Bulletin boards can do more than just swop messages. They are an easy way to exchange new computer programs between computers. While at first this sounds easy, there are some minor problems which have to be overcome. The standards give several methods. Some options have the advantage of simplicity, but are primitive, while others such as the Modem Protocol are very advanced and guarantee accuracy even if the telephone line is a little noisy. The Modem Protocol is a method devised by Ward Christensen in the United States who gave away all rights to his ideas to allow the method to become public property and is the same method used by most business computers running under CP/M. This means that bulletin boards can transfer programs between different incompatible makes of computers. There is, of course, no guarantee that after transfer they will run. It is this method, in fact that was used for transferring the text of this book for publication.

The reserved control codes are used to retain compatibility with some terminals and those allow special features which might be needed. It is by using these controls that the bulletin board can sometimes arrange for automatic log-on to save you typing in your name and password and answering various questions. The other control codes, used for stopping and starting (X-on and X-off) might be used by your terminal or computer if it can't keep up with the information from the bulletin board; or it might be used by the bulletin board to give it time to keep up with your terminal. In either case, it is usually automatic and the user does not see it.

The standards aren't formal. It is likely that as new developments and improvements come along the standards will be improved or even changed. It is hard for AFPAS to forecast what the future will bring but the standards currently in use will be retained as far as possible to keep continuity.

CHAPTER 11
Calling Direct to a Friend

Up to this stage it has been assumed that you are calling into a system that is specially set up to receive calls and provide a service, whether run by professionals or amateurs. If you want to send a message to a friend for fun or to exchange programs, however, you will have to take on all the responsibilities of normal board operators.

Earlier the difference between 'answer-only' and 'answer/originate' modems was explained. To recap, when communicating, each end of the exchange must have a different style of modem. These are called, for reference, 'answer-mode' and 'originate-mode' modems. There is no need for the originator (the person who dials the telephone number) to have the originate modem. It is conventional to work this way round, but the reverse works equally well! This type of person-to-person communication is the only time when an answer-mode modem is required unless you run a bulletin board. If you both have originate-only modems don't give up as you can communicate not by telephone but by direct wiring as explained below.

Normally, when you call into a larger system, you only see on your screen what the main system sends to you. When you see your own typing appear, this has been sent from you to the main system, and then 'echoed' back to you. This type of working is called 'full duplex'. If you wish to work in full duplex then one end of the conversation must provide this echo for both parties. The instructions for your terminal software will say if this feature is present. The system with the echo capacity must send its partner's signals back and also locally echo its own typing.

The other method is known as 'half duplex'. With this method, each side only receives the information typed from the other side. There is no need for one side to become a 'Host' and look after the other side. Each side sends its characters both to the other party and to its own screen.

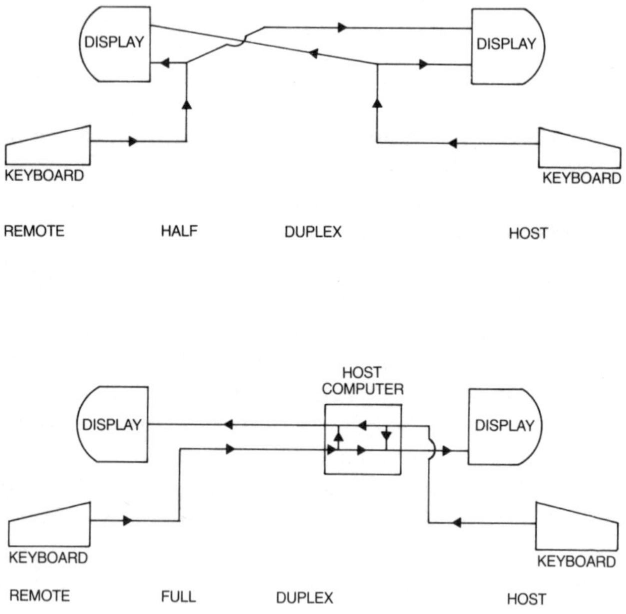

[Illustration 11.1]
Information in Half and Full Duplex

If you are close to your partner then there is no need to use a telephone at all. If directly wired modems (those which are wired electrically to the phone system) are being used, it is possible to disconnect them from the telephone line and directly wire from modem to modem. This doesn't always work as some modems need the telephone line to supply their working power but there is no need for modems at all, in fact. All a modem does is convert the data into varying tones and back from tones to data. It is perfectly possible to connect the serial lines together and transfer information directly. This is often the best way to transfer programs or files from one make of machine to another when they are otherwise quite incompatible.

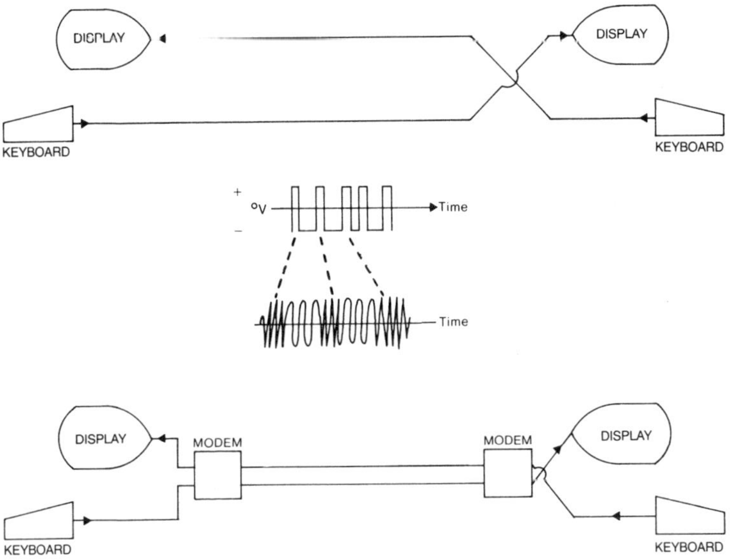

[Illustration 11.2]
Data Flows with and without a Modem

The wiring is not quite straightforward as the line that one person is sending information on becomes the line on which another receives information. This means that several of the wires of the RS-232 or RS-432 may need to be interchanged. Other small adjustments may have to be made. To handle this a special connector called a 'null modem' may be used. These are explained in the next chapter which covers the RS-232 standard.

Just as it is possible to use a modem over the telephone or directly along two wires so it is also possible to use a modem to modulate radio signals and use radio transceivers (transmitter/receiver sets) to replace the telephone line. This should not be attempted by inexperienced users and those unfamiliar with the technical side of radio. There are a number of problems to be considered including: the possibility of high voltages damaging the modem; the problem of induced radio frequency currents damaging the circuitry of the computer; and the noise generated by the computer masking any input signals to the receiver. Finally, there are restrictions on what may be transmitted under the various amateur and CB licenses.

Being able to communicate directly, whether by phone or by direct wiring, is important. It is often the only way to transfer small programs and bits of information between different types or makes of computer. This may happen if, as is the case with this book, the text is being written on one type of machine for eventual use by someone else with a different make. If you ever wish to write commercial software or games for a new machine, then this technique is also often used to allow most of the software to be written (and even tested) on an old and trusted machine before being transferred to the new machine.

Direct calling is normally straightforward, but there are some special problems that can occur. The classic problem is being unable to see your own typing. If this happens, reread the beginning of this chapter carefully. The duplex/echo combination is almost certainly wrong. The same problem can show itself as double characters on a screen. This is caused by using half duplex, as well as one end acting as a host and echo producing a second character on the screen. It must be understood that both ends of the communications must be using the same transmission speed (Baud rate) and the same details of the serial word size.

If you are wired directly from computer to computer, often it is possible to use higher speeds, and so save time. 600, 1,200 or faster Baud rates are perfectly possible. On many makes of microcomputer the highest Baud rates (usually 9,600 or 19,200 Baud) may not work perfectly even though the manual and software suggest they are functioning.

One of the most popular use for direct communications between two computers is to exchange BASIC programs. It is often thought that if two computers both run BASIC, programs should be interchangeable. This isn't always true, as the type of BASIC used may vary from one make to another; the situation is similar to different people who all speak English but use different dialects. They all understand each other but every now and then a particular word will be misunderstood. If you transfer programs you must be prepared for a number of minor changes to get everything to work.

Even transferring BASIC programs has a minor complication. Most microcomputers store BASIC programs not in the English text that you see on the screen, but in a 'compressed token' form. Here one special symbol is used for each of the keywords (LET, GO, FOR, PRINT, etc) in BASIC. Unfortunately, each make of micro uses a different series of tokens for each keyword. This means that the tokenized form of programs cannot be exchanged. It is often possible to save the program in the full text form, and this can be exchanged easily. Saving programs in this way is often called 'saving in ASCII' or 'saving in expanded mode'.

ASCII refers to the American Standard Code for Information Interchange. This is a system of coding each letter and many symbols as particular sets of zero and one bits. The code need not bother you, but remember that certain makes of machine (some Commodores, for example) do not use this code. This makes direct calls difficult to other makes of microcomputer but does not make it impossible, as some terminal programs can change the codes received and sent to overcome this problem. The terminal program will have a small built-in dictionary, and can look up the unusual code to find the standard code to translate it to.

Generally, direct computer-to-computer communications presents a few more problems than calling into a specially set up system only because there are now two ends to worry about rather than just one end trying to connect to a known system.

Many computers, small or large, do not have their own screen and keyboard, but work with terminals connected by a standard serial connector. These same terminals can be used to connect to a microcomputer to behave as the remote end of a conversation or sometimes to use the terminals in place of the microcomputer's own keyboard and display. The same rules concerning Baud rate and RS-232 connection will apply, except that a null modem is not usually needed in this mode.

Direct-wired connections modem-to-modem can often be successful for about two kilometres of wire. Connections using the RS-423 standard will allow communications over a distance of up to about 50 metres. The older RS-232 standard has a lower output and is only usable up to about 10 metres. These figures are only intended as a guide. For longer distances, special cables with amplifiers are available and work well. The need for these cables often occurs in offices when information needs to be transferred from one machine to another in a different room or on a different floor. The same thing may easily happen between two neighbours who want to communicate.

CHAPTER 12
RS-232 Standards and the Null Modem

When sending information from a computer to an outside accessory such as a modem, various methods are used for connection. Some manufacturers (fortunately a minority) use their own designs, while most use a standard method. The RS-232 standard allows data flow between various devices in a serial method – one bit at a time – between various types of devices such as a computer and a modem or a keyboard and a printer. The standard not only shows how the data will be transferred but also various specialized control signals which are often needed.

The RS-232 standard is often spoken about as a fixed standard which is easy to abide by. This is not really true. The standard was first laid down by the Electronic Industries Association in the United States and has been revised several times. The current version is known as 'Revision level C'. In Europe and much of the world the standard is known as V24, as that is the number of the CCITT recommendation that defines it.

The standard is in four parts, covering specifications for plugs and sockets, the electrical characteristics of the signals, details of each signal and which signals are used for various applications. To be strictly accurate, the V24 standard does not actually give mechanical details of the plugs and sockets but refers users to other standards. The most common plug and socket used is known as a 'D' type and usually has 25 pins for connections, although not all of these are used.

The full set of signals is very rarely used. Various groups or subsets are used for different classes of applications and some equipment may need more signals than others provide. It shouldn't be assumed that the term 'RS-232 Compatible' or 'V24 Compatible' guarantees an easy, immediate connection! The signals are of several different types: some carry actual data, some are to provide grounds or earth signals, some provide signals about the status of each end and a few are used to provide timing signals when using synchronous serial methods (which are not covered in this book).

It is easiest to describe the various signal lines in groups where each group has a function. In the standard, one end is referred to as 'Data Terminal Equipment'. For our purposes this means the terminal or computer with a keyboard. The other end is known as 'Data Communications Equipment', which means a modem to most readers. They refer to the official CCITT numbering scheme for the different wires. An appendix is given showing all the wires, their pin numbers, meanings and cross references to other numbering schemes which you might find. The arrow on each line shows the direction of the flow of information on that wire.

Group 1 – Data Transfer Only.

```
Computer                                            Modem
  DTE                                                DCE

   >------ 102 -- Signal Ground --------->
   >------ 103 -- Transmitted Data ------->
   <------ 104 -- Received Data ---------<
```

This group shows the minimum set for any communication. The 'signal ground' gives a reference level for the voltages on all the other lines. The 'transmitted data' sends data from the computer to the modem; while the 'received data' is the line on which the computer receives data from the modem. If computer-to-computer communication is being used and both had the same DTE configuration built in then a 'cross-over' would be needed as what one computer transmits is by definition what the other computer receives.

Group 2 – Control of Data Flow.

```
Computer                                              Modem
  DTE                                                  DCE

   ---- 102 ------------------------
   ---- 103 ------------------------
   ---- 104 ------------------------

 >---- 105 -- Request To Send ----------->
 <---- 106 -- Ready (Clear) for Sending -----<
 <---- 109 -- Data Line Signal (Carrier) Detect --<
```

This group adds three extra lines to allow control of the flow of data, so that data is not sent or received when one end is unprepared. The 'request to send' signal tells the modem that the computer is about to send some data. The 'ready (or 'clear') for sending' signal shows the computer that the modem is now ready to receive that same information. If the modem 8 (or other device) was not yet ready then no signal would appear here and the computer could wait so as not to lose information. The 'data line signal' allows the modem to show that it is still in contact with another modem, that is that the flow of data on from the 'data communications equipment' is still taking place. This line is normally called the 'carrier detect line' as it shows when the carrier (or signal) from a distant modem has been received.

Group 3 – Control of Interface

```
Computer                                          Modem
  DTE                                             DCE

  ---- 102 ----------------------------------
  ---- 103 ----------------------------------
  ---- 104 ----------------------------------
  ---- 105 ----------------------------------
  ---- 106 ----------------------------------
  ---- 109 ----------------------------------

<----- 107 -- Data Set Ready ---------------<
>----- 108 -- Connect DCE to Line/Data Set Ready ->
<----- 125 -- Calling Indicator ------------<
```

The three lines added in this group are to control the actual behaviour of the DCE or modem. Circuit 107 is the signal that the modem sends to the computer to show that it is ready. This is similar in concept to the Circuit 106 in Group 2. However, it is a response to the modem's status, hence its inclusion in this group. Circuit 108 is used in two different ways which depend on the type of modem or DCE being used. In the first method, called 108/1, the signal from the computer tells the modem (or DCE) to connect to the communications line (the telephone line usually). This is usually a response to the computer receiving a signal to show that the telephone line has an incoming call (i.e. the telephone is ringing). Sometimes a second method is used. In this variation, called 108/2, the signal only means that the modem *may* connect to the telephone line. Usually an extra button or switch has to be manually pushed as well. Circuit 125 is called a calling indicator or a ring detect. This line shows when an incoming ring is detected by the modem. Usually most modems don't show a steady signal on this line but allow it to pulse in time with the ringing of the telephone bell.

Other Lines

There are other lines which are used for timing purposes by

synchronous terminals. These are not covered in this book as they are rarely used by amateur communicators at low speeds. The remaining lines cover special purposes such as indicating which of several speeds should be used, whether the line condition is below standard and other functions. It is rare to find these in use outside professional installations. The appendix gives a summary of them for reference.

Nothing has been said about what the actual signals are! The standard lays down precise voltage and impedance levels which need not concern you. Briefly, the standard states that the signal must be a voltage between three and 12 volts relative to the signal ground, either positively or negatively. No signal should be in the range of zero to three volts.

The 'sense' of the signals confuses some people. The data line normally stays at a 'mark' condition. This is also used to indicate a Binary 1 or a logic function 'OFF'. When in this position the signal is negative relative to the ground.

The signal changes to positive to show either a 'space' condition which means a Binary 0 or a logic function 'ON'.

When a computer is connected to a modem it is fairly clear which end is which. The computer is normally considered to be the DTE end and the modem the DCE end. However, if the same computer is connected as a terminal into a larger computer (quite a normal thing to do) then the larger computer would consider itself the DTE end. Unfortunatcly both would now be sending signals out on the same circuit, and both would be listening on the same circuit. The result would be that nothing would work. You are more likely to meet the reverse case – a computer or terminal which was being used successfully as a terminal into a larger computer, but now won't work with a modem. Both these situations are not really faults, but just changes of use. Some computers may have a switch on them labelled 'Computer' and 'Terminal' to alter the mode.

For those without such a switch, a device called a 'null modem' may be used. This is only two connectors of the type used for RS-232 sockets connected back-to-back and wired so as to make each end see the other end as the opposite type to itself. They are available from modem suppliers at a reasonable price, or can be built at home. Two 25 pin 'D' type connectors, both female, should be connected following the wiring diagram below.

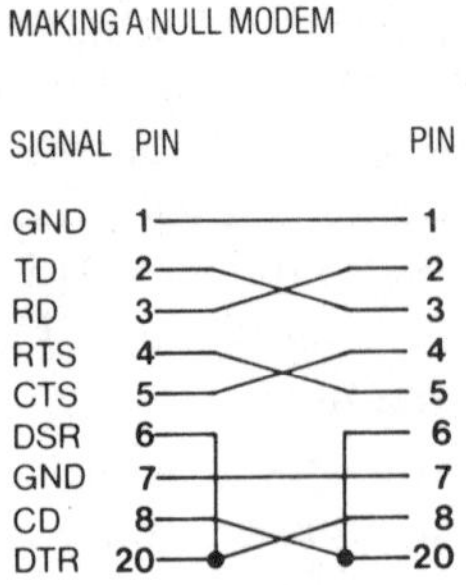

[Circuit 12.1]
Wiring Diagram for a Null Modem

Not all systems follow the V24/RS-232C standard correctly and it is useful to be able to test or experimentally cross-wire different lines. This is easily done by constructing what is called a 'break-out' box. Each of the wires from a 25-pin connector is joined to a terminal block with no connections, and the same process is repeated for another 25-pin connector. Various wires may be used from each line on one terminal block to various lines on the other, rather like the effect of a switchboard in an old manual telephone exchange. A new development is the existence of 'intelligent' cables which are able to examine the signals on the various lines and make their own cross connections with the aid of a built-in microprocessor. These are very effective, but really are only needed where equipment is being changed around frequently. For normal uses, once the settings have been found then a modified cable and plug can be made if this is needed.

Some computers do not use the conventional connectors for a V24/RS-232C interface. The most common example of this is the BBC microcomputer. If you do not have a normal connector then your computer's handbook should show which terminal has which line on your connector.

Often not all the lines are present on the computer, but the modem might need 'Circuit 105 Request to Send' or 'Circuit

108' to show a logical true or on. The modem can often be 'tricked' into working by connecting 'Circuit 106 Ready for Sending' or 'Circuit 107 Data Set Ready' to either or both.

The RS-232 standard only provides a limited amount of power, and it cannot be used for long-distance communications. However, it is perfectly suitable for normal domestic use where the wire length is not likely to exceed eight or ten metres. This is just to remind you that it is not suitable for permanent wiring across a street, for example, between two houses. The similar RS-423 standard provides more power and can be used for longer distances.

Pin	Name	DTE DCE	EIA CODE	CCITT CODE	RS232 Function / Notes on RS232 A and B and other uses
		COMPUTER MODEM			
1	FG		AA	101	Frame (Protective) Ground
2	TD	→	BA	103	Transmitted Data
3	RD	←	BB	104	Received Data
4	RTS	→	CA	105	Request to Send
5	CTS	←	CB	106	Clear to Send
6	DSR	←	CC	107	Data Set Ready
7	SG		AB	102	Signal Ground
8	DCD	←	CF	109	Received Line Signal Detect (in RS232 A and B, this is Data Carrier Detect)
9		←			Reserved for testing (positive voltage)
10		←			Reserved for testing (negative voltage)
11		←	SA		Unassigned in RS232 B and C. In RS232 A this is (SA) Supervisory Transmitted Data Bell 208A modem uses as QM equalizer mode
12	SDCD	←	SCF	122	Secondary Received Line Signal Detect Unassigned in RS232B. Is (SB) or Supervisory Received Data in RS232A. Becoming the standard for High Speed Detect in dual speed modems (103/212A, et al)
13	SCTS	←	SCB	121	Secondary Clear To Send (Unassigned in RS232 A and B)
14	STD	→	SBA	118	Secondary Transmitted Data (Unassigned in RS232 A and B) Used as (NS) or new sync signal in Bell 208A modems
15	TC	←	DB	114	Transmitter Clock (unassigned in RS232 A)
16	SRD	←	SBB	119	Secondary Received Data (unassigned in RS232 A and B. Used as (DCT) Divided Clock Transmitter in Bell 208A modem.
17	RC	←	DD	115	Receiver Clock (unassigned in RS232A)
18		←			Unassigned. Used as (DCR) Divided Clock Receiver in Bell 208A modem.
19	SSRTS	→	SCA	120	Secondary Request to Send (Unassigned in RS232 A and B)
20	DTR	→	CD	108.2	Data Terminal Ready
21	SQ	←	CG	110	Signal Quality Detector (Unassigned in RS232 A and B)
22	RI	←	CE	125	Ring Indicator
23		→	CH	111	Data Rate Selector (Direction optional)
		←	CI	112	(Code changes with direction implemented) (Unassigned in RS232 A and B)
24	ETC	→	DA	113	External Transmit Clock (Unassigned in 2)
25		→			Unassigned. Usually used as busy signal

Control Signals: + = ON
− = OFF

Data Signals: 0 (Space) = +3v to +25 volts
1 (Mark) = −3v to −25 volts

[Table 12.2]
Summary of RS232/V24 Equivalents and Uses.

CHAPTER 13
What To Do When Nothing Works

I would be happy if this chapter remained unread by everyone. If you turn here you probably have had some problems. The hardest part of curing troubles is usually finding where the trouble is, so most of this chapter is concerned with locating the trouble rather than curing it. Once you know exactly where the problem is the other chapters in this book will be able to help, unless there is a mechanical or electrical fault.

There are two different problem situations. Either nothing appears to work and you can't get through to anyone, or you can get through but the results are scrambled or faulty in some way. To simplify, diagnosis flow charts will be used. A flow chart is a diagram showing both the tests and the action you should take depending on the answers to the tests. The tests are in the diamond-shaped boxes and have two exits depending on whether the answer is yes or no. The actions are in the oblong boxes.

Some problems are so common that they are worth another look. In earlier chapters it was explained that the modems at both ends had to be matched. That means that your modem must be of the right Baud rate to communicate with your target. Not only must the Baud rate of your modem be right, but also your computer or terminal must have the same Baud rate. This must be checked. Often the Baud rates are set by the programs you are using. One end has to be set to originate and one end to answer mode. If you are calling into a large system then you should be in the originate mode.

If you are working in V21 (Prestel-type frequencies) then your software must be altered, depending whether you are in the originate mode or the answer mode – this is because the software has to send and receive at different speeds.

If the system has never worked then it is worthwhile at this stage to quickly check both the connections and the fuses. Many modems have fuses both inside their mains power supply plugs and in the modem. *Don't* put ordinary mains fuses inside the modem. Buy the correct fuses from an electronic parts shop.

Listening for a modem tone is a useful test. If you have an acoustic modem you may not be able to hear anything. However, if you blow over the cup where the earphone normally goes, you should be able to make the modem answer back with a tone. If you have a direct-coupled modem you should be able to hear the modem tones by listening in on another phone connected to the same line.

If you have got a modem tone then the next thing is to repeat the test but this time pressing a few keys from your keyboard. With each key pressed, the tone should alter for a fraction of a second – making a tone like a 'chirp'. The presence of a chirp means that your information is getting through to the modem.

If there is no 'chirp' sound then the problem might be happening before the modem. Set your terminal program to full duplex and for the same speed (300 Baud for preference, not Prestel) for both directions. With the RS-232 cable unplugged from the modem there should be nothing on the screen when you press a key. If pins 2 and 3 of the RS-232 Connector (the 'transmit' and 'receive data' lines) are connected together, then whatever you type should appear on the screen. If nothing appears, then you may have either a faulty cable, a faulty computer or RS-232, or possibly software that is not working. Check the software *very* carefully if you typed it in yourself. If you bought your software then you should now ask your dealer for advice.

If the last test worked, the problem must be in the modem or the connections to it. Common problems here include the transmit and receive lines being wrongly connected so that

Complete Failure Flowchart

ILLUS 13.1
FAULT FINDING WHEN
THE MODEM APPEARS
QUITE DEAD

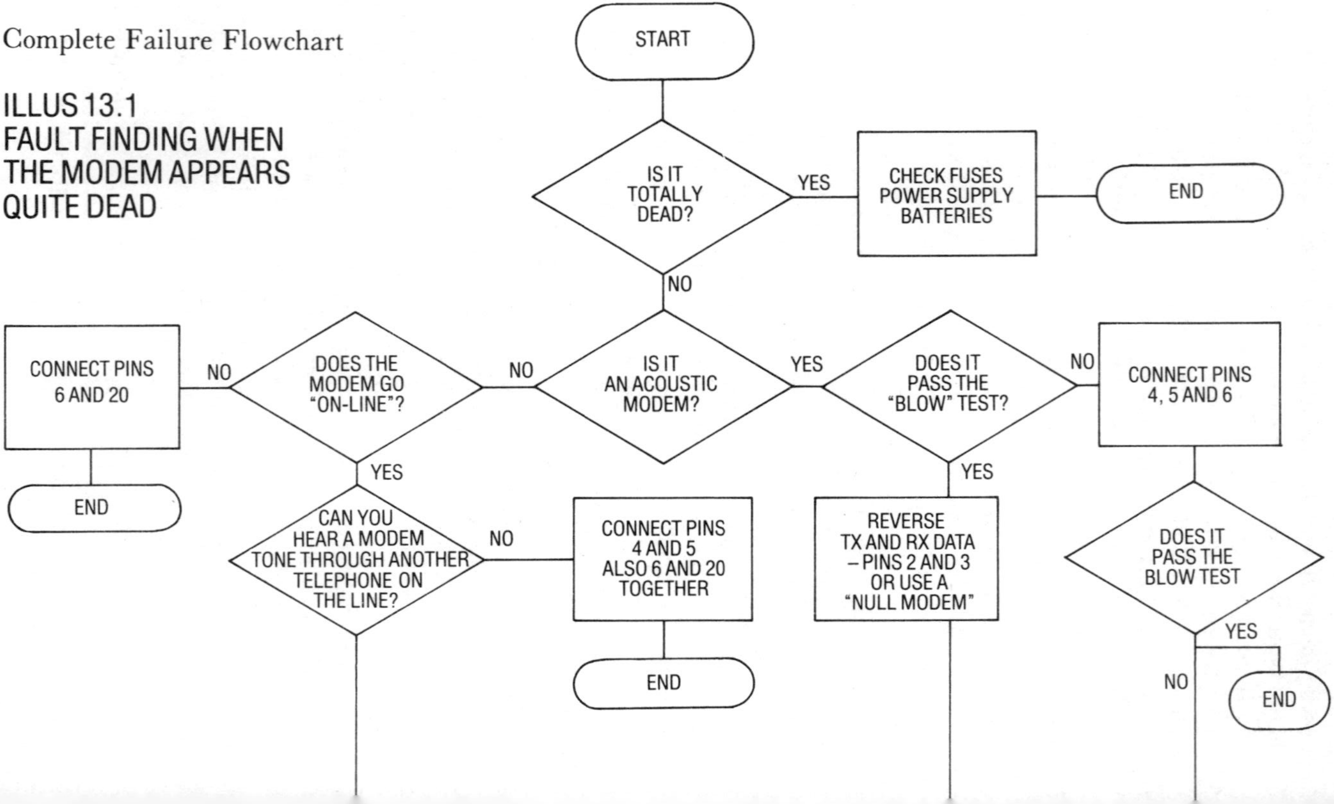

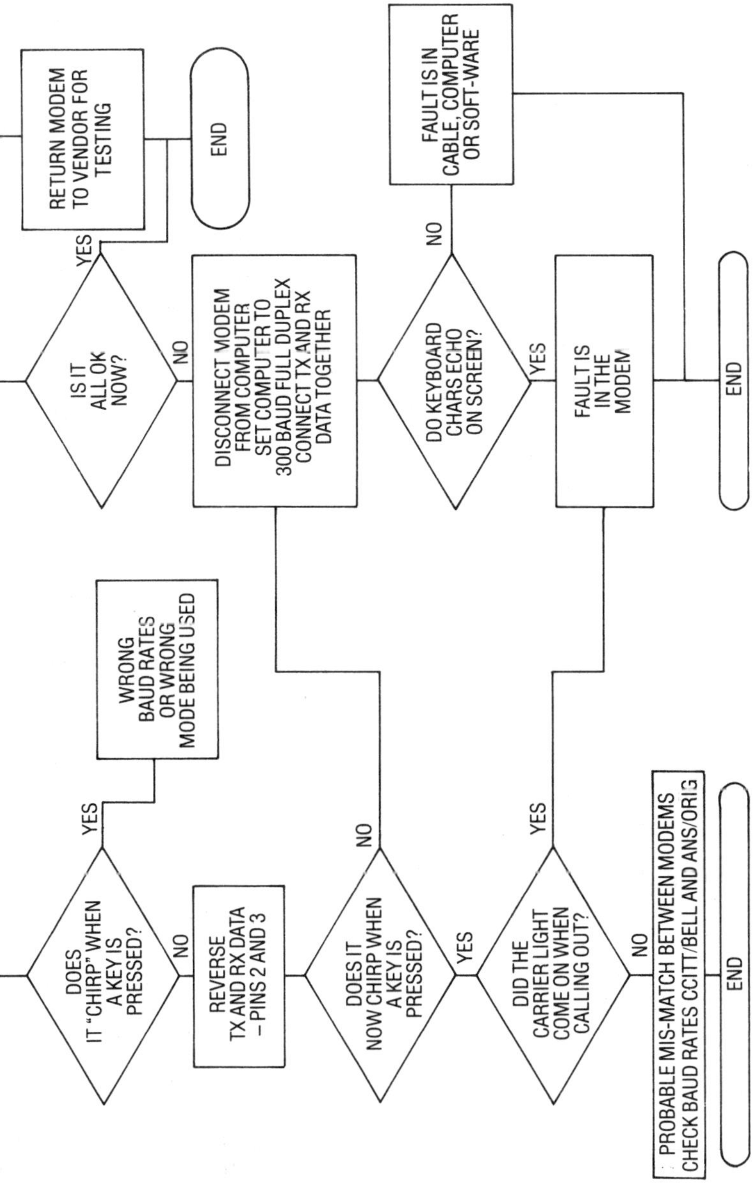
DOES IT "CHIRP" WHEN A KEY IS PRESSED?
YES
NO
WRONG BAUD RATES OR WRONG MODE BEING USED
REVERSE TX AND RX DATA – PINS 2 AND 3
DOES IT NOW CHIRP WHEN A KEY IS PRESSED?
NO
YES
DID THE CARRIER LIGHT COME ON WHEN CALLING OUT?
YES
NO
PROBABLE MIS-MATCH BETWEEN MODEMS CHECK BAUD RATES CCITT/BELL AND ANS/ORIG
END
IS IT ALL OK NOW?
YES
NO
RETURN MODEM TO VENDOR FOR TESTING
END
DISCONNECT MODEM FROM COMPUTER SET COMPUTER TO 300 BAUD FULL DUPLEX CONNECT TX AND RX DATA TOGETHER
DO KEYBOARD CHARS ECHO ON SCREEN?
NO
YES
FAULT IS IN CABLE, COMPUTER OR SOFT-WARE
FAULT IS IN THE MODEM
END

[Diagram 13.1]

Some Results Flowchart

ILLUS. 13.2
FAULT FINDING –
THE MODEM SEEMS
TO WORK

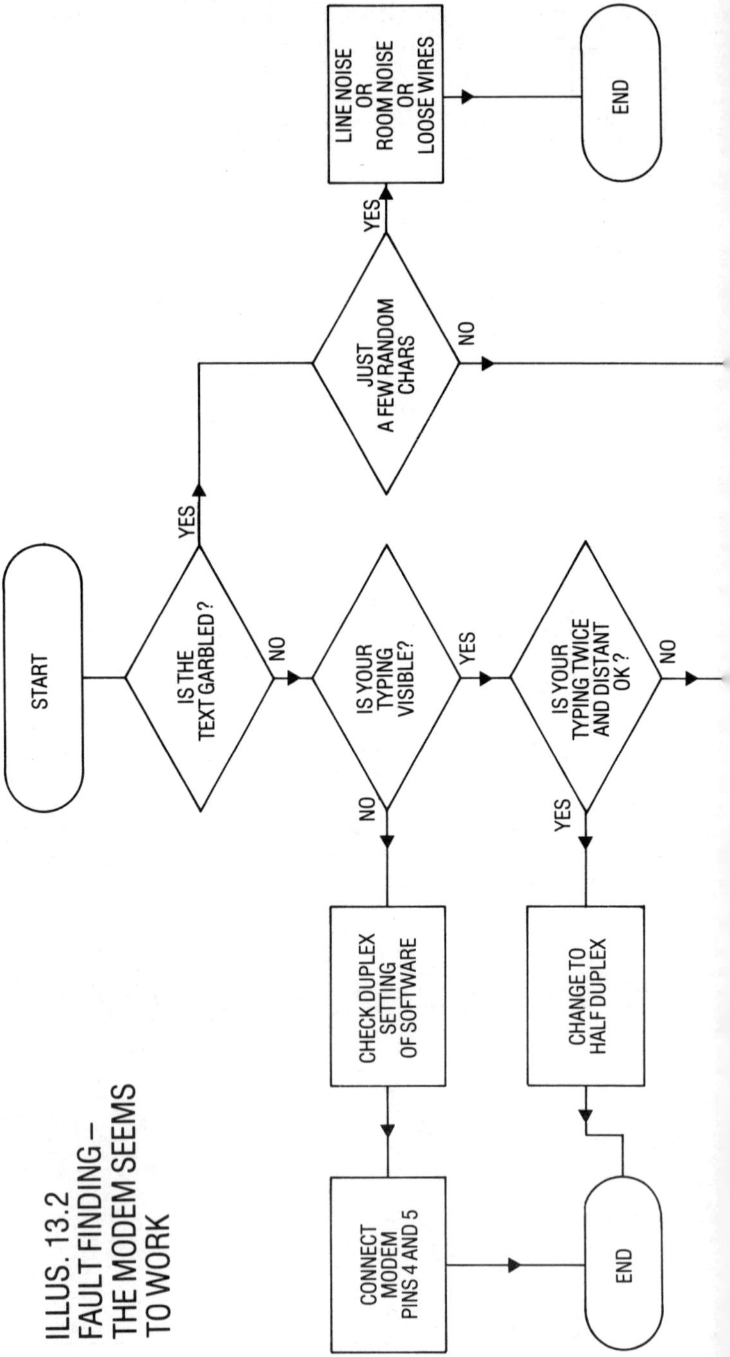

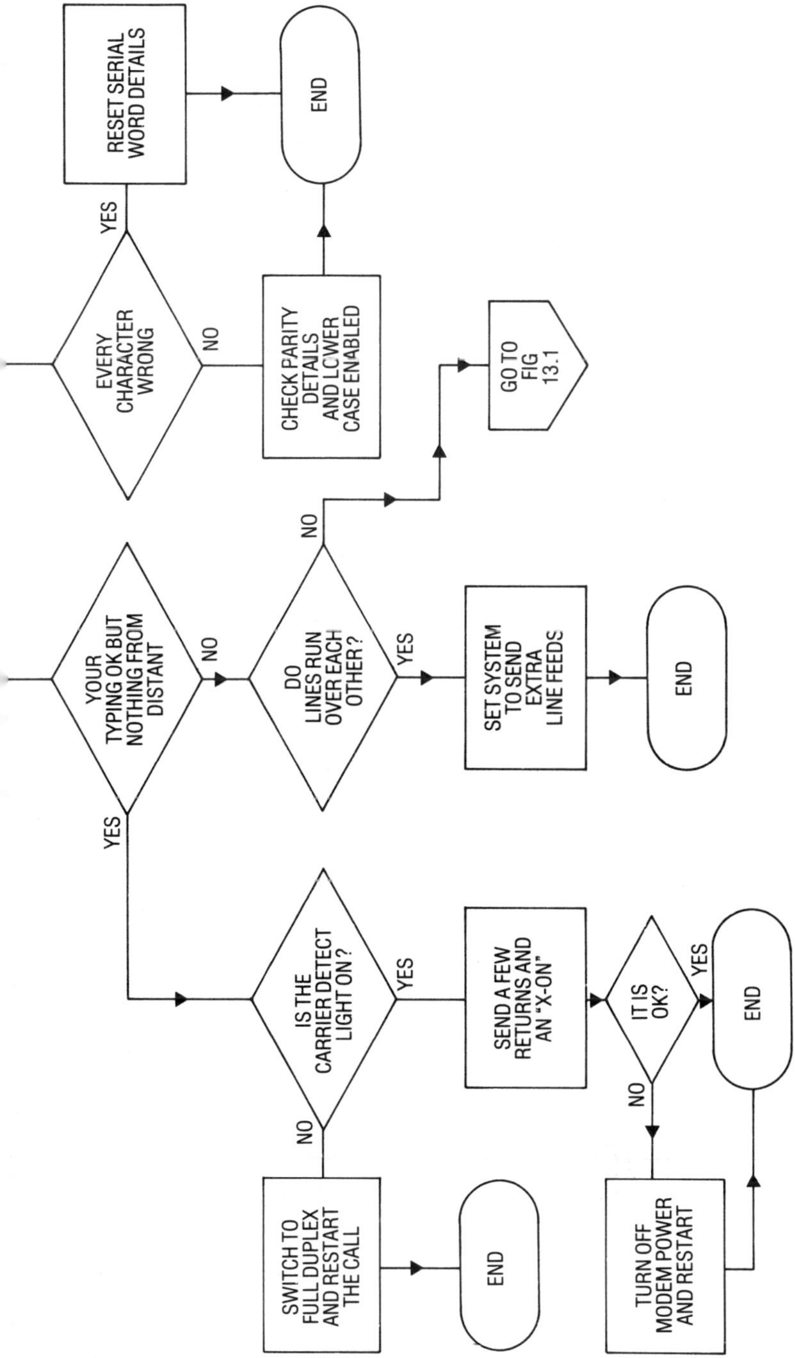
EVERY CHARACTER WRONG
YES
RESET SERIAL WORD DETAILS
END
NO
CHECK PARITY DETAILS AND LOWER CASE ENABLED
END
YOUR TYPING OK BUT NOTHING FROM DISTANT
NO
DO LINES RUN OVER EACH OTHER?
NO
GO TO FIG 13.1
YES
SET SYSTEM TO SEND EXTRA LINE FEEDS
END
YES
IS THE CARRIER DETECT LIGHT ON?
YES
SEND A FEW RETURNS AND AN "X-ON"
IT IS OK?
YES
END
NO
TURN OFF MODEM POWER AND RESTART
NO
SWITCH TO FULL DUPLEX AND RESTART THE CALL
END

[Diagram 13.2]

they need interchanging. A null modem can be worth trying. Traditionally, a 'breakout box', as described in the previous chapter, is used. This will allow you to experiment a little without taking more than a few minutes.

If the modem is making no sound at all, the first thing to check is that any lights that are fitted and should come on do. If they don't and the power supply and fuses are correct then the modem may be faulty and should be returned to your supplier for repairs.

If the lights come on but there is no sound it is possible that the modem requires the 'Data Terminal Ready Signal' on pin 20 of the connector (Circuit 108). Some computers cannot send this signal out. On other computers this signal can be sent, but only if the software specifically sends it. The same is also true of the 'Request to send' signal on pin 4 (Circuit 105). In both cases, the missing signals can be imitated by a positive voltage. This is usually easiest to obtain for the modem itself. Test pins 5 or 6 (Circuits 106 or 107) for a positive voltage relative to the ground on pin 7. If there is a positive voltage then connect that pin to either (or both of) pins 4 and 20.

The reverse is also possible, but is rarer. The modem is present and working but the computer will not send any information to it and ignores it. The modem will make a sound but does not chirp when the keys are pressed, and the modem's 'Carrier detect' light will come on when you call another system even though no contact can be made. The solution is to provide the possibly missing positive voltages on pins 5 and 6 ('ready for sending' Circuit 106 and 'data set ready' Circuit 107). The voltages are usually best taken from pins 4 or 20.

One of the most common situations can hardly be called a fault, as it is a design feature which leads to incompatibility. The computer on which you are typing has an RS-232C interface, but no clue is usually given as to whether the interface is designed to use the computer as a terminal into a larger system (the DTE end) or as the larger system which a smaller terminal can plug into (the DCE end). If you examine the previous chapter's wiring diagrams it is possible to work out that if a DCE-type computer is connected to a modem then nothing will work! The cure for this is a small device called a 'null modem'. This can easily be built following the diagram in the previous chapter.

If you have got this far, your modem should be working. If you are still not getting the results you expected, for example, if everything that you type appears on the screen twice:
HHAALLOO TTHHIISS IISS WWHHAATT CCAANN HHAAPPPPEENN..
then the problem is probably that your modem or terminal is set to half duplex when it should be set to full duplex. You have been sending information to the screen as well as receiving the same information from the host that you are calling.

The reverse situation will mean that you never see your typing on the screen, but the host will appear clearly. This is usually caused by running in full duplex when you should be in half duplex. If you are just calling a friend and both of you are in full duplex then one of you (traditionally the one with the modem set to answer frequencies) should also be producing an echo for the other's screen.

Strange symbols on the screen are usually caused by having the wrong settings for the word size on your serial output. If this can be adjusted try setting it to a seven-bit word with one start bit, one stop bit and even parity enabled. This will almost always work.

The occasional odd symbol can be caused by having the parity set wrongly. Many computer systems show a character received where the parity (an error check) doesn't appear correctly as a special character, such as a white space, on the screen. If the symbols are varying all the time, the problem is likely to be noise on the line causing the incoming sounds to be misheard and consequently the information to be corrupted.

A few users have found that when making long-distance intercontinental calls that the line keeps on going dead. This is caused by the telephone companies fitting special devices on the line to keep speech clear, and stop echoes. There is no easy solution for this, but some modems are able to send a special tone to switch off this equipment and give a better call.

CHAPTER 14

Safety – Yours and Your Equipment's

It is worrying that most computer hardware books ignore this topic completely. Many professionals happily forget that not all voltages are very low and comparatively safe. There are three main danger points to be considered: your own safety and the risk of electrocution; your equipment's safety and the risk of damage; and, finally, the risk to others working elsewhere, such as telephone engineers.

Unreliability in computers can often be traced to badly wired mains plugs. You will find the details and colour coding of how to wire up a plug written with each new plug you get.

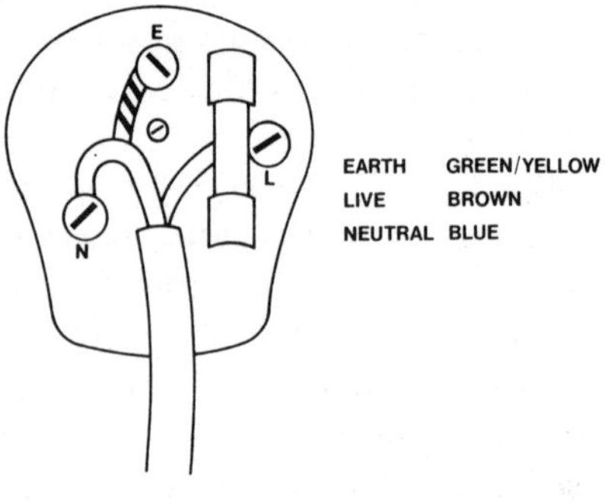

[Illustration 14.1]
13-Amp Plug Wiring Diagram

It is important when wiring a plug to make sure that the wires go to the right terminals – if you aren't sure then ask for help. The strain release clamp which holds the wire in place is very important. With constant use and movement the wires can begin to come loose from the terminals. At worst this can cause a short circuit and a fire. At best (and still disastrous) it will cause intermittent errors on your computer which can be hard to trace. Worn mains wires should be thrown away and replaced. There is never any excuse for using chafed or worn wires, or allowing the outer covering of a wire not to be gripped properly.

The fuse supplied when you buy a modern plug usually has a 13-amp capacity. This is the amount of current it can pass. Your computer is unlikely to draw more than two amps at maximum. You should take out the 13-amp fuse and replace it with the correct fuse for the system. If you are not sure what fuse is needed just add up all the power consumptions (in watts) of each part of your system. The fuse needed will be found by dividing the power (in watts) by 240. Fuses are normally available for mains plugs in 1,2,5 and 13-amp capacities.

Not all electric power is the same. The measure of the strength of the 'push' in an electric circuit is given by the voltage. This is the difference between two levels. If no base level is mentioned then the measurements are taken relative to the 'ground' lead, which is a neutral point. Items designed for different voltages cannot be used on the wrong voltage power supply. This is especially true of equipment coming from the United States where the common voltage is 110 volts where we normally use 240 volts.

A transformer can be used for changing the voltage level of mains-powered equipment. It is generally suitable for use in supplying non-mechanical items. Extra care must be taken with mechanical items as many are also sensitive to the frequency used in the alternating current mains supply. You should always consult the manufacturer before using American specification equipment on European power supplies.

The fuses used inside electronic equipment are not the same as those fitted in plugs. The circuits used are more delicate and can be quickly damaged. Two special types of fuses were invented for this - fast and slow-blow fuses. The fast-blow fuses are designed to break the current as quickly as possible if it ever reaches the maximum current, even for a short time. Slow-blow fuses are meant to allow small fast surges of current through without breaking the circuit, but they will work if the current stays at the higher level. The equipment designer will have chosen the fuses knowing exactly what loads and surges are possible. If a fuse blows look at it and check it. You must replace it with a fuse of the same value current and same type.

Many electrical and computer notes mention the words 'earth' and 'ground'. These both mean the same thing and refer to a neutral point that is eventually connected to the earth or soil in the ground. This does not mean that a wire pushed through a window into a garden is satisfactory. If you have any doubts consult a qualified electrician. Never try and use gas pipes as ground terminals. In these days of plastic pipes the water pipes should not be used either. Some electrical equipment is not designed to be grounded and just runs with a two-wire cable to the mains. You can't connect other items to a terminal on such equipment marked 'ground', and so provide a ground for your modem. It is important to make sure that your system has a real ground at some point and that other parts are connected to it.

While on this subject some users of 110 volt equipment use an auto-transformer to reduce the European 240 mains to 110 volts. These transformers do *not* provide the normal isolation that most transformers give. Don't assume that because a transformer is being used that the system is isolated from the mains.

The danger of mains voltage is usually not to the user of the computer, but to anyone working on the telephone lines if you have connected your equipment up badly. It is very easy to use an auto-transformer and not realize that the entire system is working at a high voltage. No problems will be apparent but when the system is connected to the telephone system any repairmen could be electrocuted.

Your computer is also vulnerable to the reverse situation. Much of the circuitry is very sensitive and can easily be damaged by very short bursts of high voltage, even if they only last for a thousandth of a second. Your modem should have built-in protection against lightning. It is easy for high voltages to pass through the telephone lines into your modem and on to your computer. The best advice that can be given is to disconnect your modem and computer from both the telephone and the mains when they are not in use. These same spikes of very high voltage can be found on the mains power supply. They are more common in country areas, and where cables are not run underground, but they can happen anywhere. Precautions against these short spikes of power can easily be obtained by fitting special plugs or boxes which contain protective circuits. They are cheap and nearly always worthwhile.

Safety is generally a matter of knowing and understanding what should be done. There is nothing wrong in going to your dealer and asking how a particular problem should be overcome. It is not a clever idea to 'botch' a job. The signal earth lines on the RS-232-C should not be used as a power supply earth. Cables should not be stretched so tight that the slightest pull will dislodge them. Long mains cables are a serious hazard. They usually come loose at some point and cause the system to crash because of the lost power. Worse still, people trip over them when they are in a hurry.

Liquids are the biggest single hazard to microcomputer installations. No drinks should be allowed anywhere near your computer. Apart from the dangers of spillages damaging keyboards and other parts, liquids can cause short circuits and fires inside monitors and televisions. *Never* stand cups of coffee on top of a monitor or VDU.

CHAPTER 15

Running Your Own Bulletin Board

More fun can be had from running your own bulletin board than from any other area of communications. Most system operators or 'sysops' are converted callers who have enjoyed the services that others have provided and thought (often rightly) that they can do as well or better. This can be strange at first. As a sysop I have chatted with many regular users, whom I now know as friends but I have only personally met a few of these guests. The bulletin board has become my own telephone answering machine. There is also the satisfaction of being able to create and organize a computer system which could actually do something.

Most microcomputer users are micro-holics or addicts of the keyboard. The symptoms are clear and well known: a constant craving for a new program, and a habit of staying up until the early hours of the morning to work on the computer. Running a bulletin board is the safest cure for this disease. The sysop has all the pleasure of using and creating a computer system without the constant need to be there typing at the keyboard. Like a good garden, the system will run itself happily, just needing the occasional bit of attention.

The system operator has certain duties. He must remove messages and text which have become old or boring. He must cut out sections of the bulletin board which have no interest and encourage other areas which do.

To run a bulletin board needs a little more equipment than is required to be a caller. The essential feature is that your modem must be able to answer by itself and have an answer mode. Obviously a telephone is essential. This is best not shared with others in the household. Experience shows that the use can be very high, and no one likes to be constantly disturbed by people who want to speak to your computer.

It is possible to share a telephone line between human and computer users by using the ringback method. In this method the computer does not normally answer the telephone and ordinary callers can phone in as usual. Callers who want the computer must let the phone ring twice and then hang up. They then phone back within half a minute and the computer will answer them. This method is not foolproof. Some voice callers will be disturbed by the unpleasant modem tones; others will not be able to get through because the telephone will often be in use for computer calls. However, the method does allow a bulletin board to be run without the extra cost of a second telephone line.

A separate telephone line is the best alternative as it allows a 24 hour service without disturbing the normal use on the main phone line. Contact your telephone sales office and tell them the purpose of your second line. Given prior warning they are usually most helpful.

If the system is to work unattended then the modem must be able to answer by itself. Most auto-answer modems are quite reliable. However, you should check that the type of auto-answer is compatible with the software you are going to use. (The two different methods of auto-answering are explained in Chapter six).

It is almost essential to have some sort of disk storage. There have been a few simple bulletin boards written for some types of microcomputers which were entirely in memory, but they were only intended as demonstrations. Usually several floppy disk drives are used, and often a hard disk which can store several million characters. It is quite possible to operate a bulletin board with just one floppy disk drive. Usually this means sacrificing many possible features and concentrating on just the messages in the system.

To run a successful bulletin board two features are essential: regularity and reliability. Your callers must know that at the agreed times the board will be there ready to receive their call. Systems which run at odd hours, or where the operator switches on only on request never gain popularity. The callers expect to see a bulletin board running just as they expect the telephone operator to be at his position. The fact that a bulletin board might be run by an enthusiast on a very small system will not be known by the callers.

Reliability is even more important. A bulletin board normally has to run totally unattended. The people using it will often have no technical knowledge at all. The entire system must be very reliable so that it does not break down while unattended. The software must be written so that if a wrong entry is made or a fault is found then the system can recover by itself. Special care has to be taken to make sure that people who hang up in the middle of call don't leave the system stuck waiting for a key to be pressed, or damage any information by leaving any files of text open.

Regular maintenance is essential for an interesting board. The old messages should be cleared out every few days. To organize the maintenance it is usual to take a daily spare copy of any files which change all the time. This is called an archive copy and is used if the normal rota is muddled. The main files of messages are cleared and then the entire disk with the information is copied to another disk. The best method of doing this is called the grandfather-father-son method. Three disks are used for each disk of data. When a disk is copied for security it is copied from the father to the son. The grandfather disk contains the information from the backup before. When the process is next repeated the old grandfather disk is used as the new son; the old father becomes the new grandfather, and the old son becomes the new father. This is repeated in a simple cycle. This method means that the previous information is always there and if anything should go wrong not all the information is lost. It is a little muddling and so most users keep a record card showing which disk is which. For example, if there are three disks A, B and C:

Date	Grandfather	Father	Son	Unused
1			A	B,C
2		A	B	C
3	A	B	C	
4	B	C	A	
5	C	A	B	
6	A	B	C	

Organizing a bulletin board need not be a single-person activity. Many very good boards are operated by clubs or by two individuals who share the work. A typical bulletin board with only one telephone line can take about 100 calls a day if the maximum time is half an hour, but many callers only stay on for a few minutes. The use of the board tends to be self-limiting. If too many people are using the board then it becomes very difficult to get in and so many users won't use it but will try elsewhere. In a typical system the sysop might spend 10 minutes every other day reading his mail, and about two hours once each week to clear up old and unwanted messages. If the sysop writes editorial articles (and nearly everyone does) then this, of course, will take more time.

Generally, it is great help if you can enlist others to help you run a bulletin board. It adds to the interest greatly if a number of people regularly write articles, reviews and other topical features. A bulletin board should be a two-way communication medium and it is important to encourage all the users, no matter how inexperienced, to leave some kind of message for others after each call.

It is usually best to use a pseudonym if you wish to run a bulletin board system. There are always a few eccentric users who wish to visit systems unannounced, and at unsociable hours. Anonymity is the sysop's only protection against this type of behaviour, but it is difficult to remember a false name and use it consistently.

Running your own bulletin board is enjoyable, but it does involve a great deal of work and expense. Sponsorship by a commercial company is the ideal answer towards the running expenses but it is not always possible. If you belong to computer club then you might like to consider setting up the venture jointly. Not only does this spread the cost, but you can also spread the workload. Most systems will allow all the housekeeping to be done remotely. This means that several people can look after a bulletin board, even though they don't often meet together in person.

When a bulletin board is designed, you should remember the educational aspects. Try to have small areas where people can exchange messages for public reading, send each other electronic mail and upload and download software. Other things that can make a bulletin board interesting include surveys of the opinions of callers and their display; remotely run games and amusements; advice pages run by a doctor; lonely hearts column; weekly diaries of events; buy and sell columns, job shops. The easiest way to find new ideas is to look at a good local newspaper.

CHAPTER 16

Software Available for Bulletin Boards

There is no reason at all for not writing your own software to run a bulletin board; but there seems little point in re-inventing the wheel when many excellent systems are available at a low cost. Most types of bulletin board software allow you quite a wide scope to change the system and alter it to your wants. All the software commercially available works well. The differences lie in the amount of freedom which you are given to change and design the system. At one end of the scale lie the TBBS family of systems where complete freedom is given and the system can imitate the behaviour of other packages or be extended easily; at the other extreme are systems such as the CBBS family for CP/M where all boards retain a close similarity in organization.

Many of the packages originated in the United States. This is partly because the telephone charge system there makes local calls to a bulletin board very economic. There is no problem in using these systems in Europe. The software itself does not know whether it is working in CCITT or Bell standard frequencies. All the software packages will run at 300 Baud without problems, but 1200 Baud is not always successful. Very few packages are available for more than one machine make. If you have one of the less popular makes of microcomputer you may find that you have no choice about the software package that you will use.

Users of Apples, Tandys, CP/M and IBM PC microcomputers will have a larger range to choose from. Historically Tandy computers have always been used a great deal as bulletin boards because they are one of the cheapest range of machines to provide a full RS-232-C interface. Some microcomputers only provide a restricted RS-232-C interface with only the Group 1 data lines present.

To give an idea of the range available it is worth looking at some of the common types of bulletin board available. One of the oldest and most comprehensive types is the Forum-80 System written by Bill Abney in the United States . This is written for the Tandy TRS-80 range of microcomputers and has shown itself to be reliable and popular. The system consists of a large number of programs which are nearly all written in BASIC. This shows itself in the slight delays involved when the system is running.

Like all the systems available, Forum-80 offers public bulletin board communications between users, and a private mail facility. Forum-80 is particularly easy for the operator to maintain as the messages may be set to delete themselves after a certain date or after they have been received. As a large part of the sysop's time is taken up with this kind of maintenance any automatic feature like that is a great time-saver. The messages are not actually deleted at that time, just marked as deleted, but they disappear from the user's view. They are actually erased only when the sysop chooses to run the special cleaning up programs.

Forum-80 is unusual in having a list of indecent words which it will censor. If a message includes one of these words, special action can be taken. The system is reliable and its menus and selections can be changed by editing the BASIC programs supplied. The system really needs a minimum of two double density disk drives (about 300K of storage) to work. Forum-80 includes the facility for users to string commands together to go direct to a given function. This allows users who call in regularly for one particular feature to get a very good service.

TBBS or The Bread Board System is a new bulletin board system which appeared originally for the Tandy TRS-80 range. It is now also available for the IBM PC, the Kaypro 4 and the Epson QX-10. It is an unusually flexible system. Unlike Forum-

80, the software consists of a small interpreter which reads menus constructed with a menu editor. There are over 60 different kinds of commands which can occur in a menu, the form of which is entirely up to the sysop. The major system options are also readily changeable and are kept in a separate file. The system can have any combination of facilities. There is no limit to the number of menus that can be accessed. Commands are given to go to a menu and then return to the present position (an equivalent of GOSUB) or just to go to a new menu (a GOTO).

This flexibility gives some new operators a difficult time. Specimen boards are supplied but it takes several months to realize the potential of the system. There is no automatic deletion of old or received messages, but callers can be given a reminder to delete the messages after receipt. The system is unusual in giving very good support for questionnaires and surveys which may be conducted on-line by callers. TBBS may be run entirely by a remote sysop who accesses the system by a modem. This flexibility includes changing all the configurations and menus as well as editing message and user-log files. The system allows up to 25 separate message areas and 255 levels of user authority. Different users can be made to see different parts of the board. Similarly, different people can be subsidiary system operators for particular areas. Callers may have authority only to read or only to write in particular areas.

TBBS gives excellent message-searching features, allowing searching in forward or reverse order, as well as searching for messages that you have previously marked as interesting. It has a very powerful feature of allowing messages in a chain, where one message is a reply to previous one, to be read in an automatic sequence. This makes it easy to follow conversations and trains of thought. The outside caller normally finds this to be a particularly valuable aid to understanding what can be a lot of separate short messages in the files.

TBBS is a deluxe system, but it does require more effort than most to get it running to its proper capacity. The system can run with only about 150K of storage but, as with all boards, more storage makes the running much easier.

CBBS or Computer Bulletin Board System is a popular system for CP/M-based computers. The software is very reasonably priced and powerful. Unfortunately, because of the differences between different CP/M computers, there may have to be a little work customizing CBBS to get it to work on your particular micro. This needs a knowledge of the hardware of your computer and of assembly language. Once this work has been done the system is effective and well considered.

CBBS was designed by Randy Suess some years ago and has been regularly upgraded to provide more and better facilities. It is less easy for beginners to use than the two systems already described. Menus and help files are present, but the normal position is for little or no help and prompting to be given. This is fine for experienced users and saves considerable call time. Like Forum-80, CBBS allows users to string commands together in one chain. In fact the system relies upon this to build up specialist commands.

CBBS does not allow the selection of messages in reverse order, or conversations to be followed through but it provides excellent searching on either the inclusion or the exclusion of keywords.

There are software packages available for most micro-computers to run bulletin boards. If you want to buy a package you should consider the following questions:

What happens if a caller just hangs up in the middle of a call?
What happens if a caller stays connected but doesn't enter anything for 15 minutes?
Can different classes of user be excluded from some areas?
Are private messages allowed?
How often does the system have to be 'cleared' of old messages?
What limits are there on the number of users and of messages?
If a fault occurs is there a log kept for the system operator?
Can users break out of the system and get access to the computer?
Do users have passwords to preserve security?
Can the operator interrupt to talk to a user?
Can programs in BASIC be run remotely?
Can the system be set to ask questions and record the answers?
Is there a provision for information and help files for new users?

After you have used other bulletin boards you will find specific features that you might like to include on your own set up. Some systems allow this, others don't. In general, the security of a system, and its recovery from callers who hang up or who don't have a proper carrier is the most important thing.

Most bulletin boards are really only programs running locally on a computer, which have also been modified to allow input and output through a modem. This means that writing your own program is not as hard as it might appear. If you want to write your own bulletin board you will find that there are four main parts to be worked out. The main menu program, the files, the drivers and the interrupt system.

The main menu program is the the outside shell which callers see. This is usually a program which asks questions of the user, and then carries out the appropriate action. To write a program to do this you should ignorc thc fact that input and output is via the modem, and start off with input from the keyboard and output to your screen. You will need to remember that not everyone has the same width screen as your system. You will find yourself having a program that branches to subroutines depending on menu choices. Don't be too worried if the program gets large as it is often possible to change a section from a subroutine to a separate program.

The files are the core of any system. Usually two main files are kept for messages. All the message text is kept in one continuous file, sometimes without any marker to separate the different messages. A separate file is also kept which has all the details of the messages. Typically this file would contain who the message was from and to whom it should go, the date, and any password to read the message. This file would also include a record of where the message was in the text file. This detail file is often called a 'header' file and is usually in records of a fixed size. This means that each header can be easily accessed and used to work out where the text is and what should be done with it. There are usually separate files to keep details for each user, such as when he last called and what his personal password is. If you have information or help texts, these too would be kept as separate files so that they can easily be changed without altering the main program.

The drivers are the key to making the system work. Without the drivers the system will work only for those who sit down at the keyboard in front of the computer. The drivers' job is to listen for input from either the keyboard or the modem and put both into the main program as input. Similarly, everything that is sent out by the program should be sent to both the screen and the modem. You must make sure that everything that the distant end sends to you through the modem is echoed back to the distant end. In a full duplex system the callers will only see what is sent by the host computer, not what they type themselves. On some computers the drivers are already written and are part of the operating system.

The final requirement of a bulletin board is some way to check for accidents as they happen. The most common problem is for a caller to hang up or for a telephone fault to cut short a call. When this happens the bulletin board must be able to stop, shut down and restart ready for another caller. This is most usually done by using 'interrupts'. The computer is able to stop its normal actions and branch to a separate program, either at frequent short time intervals, or when an outside signal calls for help. This is used to watch the 'carrier detect' signal from the modem and check that the caller is still there.

It is fairly easy to write your own system, and if you have very little storage space it can be worthwhile. The commercial packages are written assuming two or three floppy disk drives. If you write your own you can pare down everything to fit on one disk drive, or even to make a system which stores everything in memory and has no disk drives. Most people find, however, that in the end it is worthwhile buying a professionally written piece of software, just to save the time in development, and get on with enjoying themselves.

CHAPTER 17

Calling The World and Beyond the Fringe

The telephone system spans the world both with direct dialling and with operator help. The cheap off-peak rates for calls allow you to make calls to other countries without spending a vast amount. Take care however, as the cost can still mount up quickly. Many other countries offer interesting services which are free for all to use.

The world uses two main groups of standards – the CCITT and Bell frequencies. In general, CCITT is used throughout Europe, Scandinavia, the Middle East (with amateurs – officially it's Bell there), Australia and South Africa. The Bell standards are used in the United States, Canada, Central America and parts of the Far East. To call to a Bell area you will almost certainly have to use a Bell standard modem. Very few systems have the ability to accept calls from a different standard. Where this is a possibility it is only by agreement with the system operator. The details of Baud rates and serial word size given earlier (300 Baud, seven data bits, one stop bit and even parity) are fairly universal no matter where you call.

Nearly all countries also use a full duplex method of transmission. All your normal software and experience of work in Britain will be applicable elsewhere in the world.

Most countries have boards running in their own language only. Some boards in Scandinavia, however, include areas with English commands and English help details. The largest number of bulletin boards are in the United States where the bulletin board movement has been going well for a number of years.

Good manners when calling foreign boards is essential. It will help the system operator if you don't fill in a registration form unless you intend to call the board again. Most systems have a limited storage for user details and cannot afford to have thousands of unused registrations from around the world.

Don't just be a reader. Leave a message for other users, even if it is only greeting from a distant country. The bulletin board only survives because of the messages on it. People will be interested to know how you got the board's phone number, what kind of equipment you use and how the board compares with those you are used to.

Don't abuse foreign (or home) boards. Try to avoid using false names and never leave messages about politics or religion. Bulletin boards are meant as a friendly place to chat and to exchange information, and there are always a few people who may take offence to personal opinions.

One of the hardest things when calling a foreign language board is to get help or find your way out. If you don't know what to do try pressing the carriage return (also called 'Return', 'CR/LF' or 'New-line' on some machines). This will usually take you out of the area you are in. A question mark is universally understood as a request for information and most systems will give you some reply. To exit a system try using T, Q or E (Terminate, Quit or Exit), symbols that are used in most countries.

If you have a Bell modem the headquarters boards of the common software systems in the United States all contain a list of other boards running the same software. Try phoning:

TBBS Headquarters (0101)303-6904566
CBBS Headquarters (0101)312-5458086
Forum-80 Headquarters (0101)816-8617040

The term 'hacking' has changed meaning several times since

microcomputers first appeared. Originally, as far as I can tell, 'hacker' was used to describe a person who assembled his own equipment and designed the supporting software. This changed fairly quickly to refer to someone who was would modify, alter or customize software. For many years to be called a 'hacker' was a mark of pride for an amateur as it suggested a competence at a professional level for modifying and adjusting software, usually in machine code. With the coming of modems and computer communications, a hacker came to mean a person who made calls into other systems. In some areas it specifically refers to those who call and succeed in entering systems where they have no authority to go. With the increase of governmentally and commercially distributed computing it is common for secondary access to be allowed from ordinary telephone numbers, and this is the common entry point for most hackers. Nearly all systems are provided with good password protection. All the examples of hacking where illicit or unauthorized entry has taken place of which I have ever heard have actually started with a leakage of a password or a phone number from an authorized user.

The hacker is a modern phenomenon and the result of an anti-authoritarian approach to the world. At first sight the hacker may seem to be a harmless nuisance. Few hackers have got or even want to go beyond the stage of just seeing that a system is there. The presence of hackers in a system should be taken as a warning that the system security is weak and could be penetrated by those with more deliberate intentions. Experience has shown, however, that elaborate security precautions will only encourage hackers to spend a great deal of effort and ingenuity overcoming them.

Many modern systems are being fitted with secondary detectors which stop the modem from answering unless the correct dialling numbers are sent after having dialled in. This is probably fruitless. If a number has been given away, so probably has any secondary number. Good password control and the devalidation of passwords when users leave is the best security available. The problems lie usually with the users and not the equipment.

This type of hacking is fundamentally anti-social. Callers have reached and used main serious users such as the university computing networks, financial and news systems as well as several banks. There is little fun to be had from these systems, and the few people who have done this have caused great difficulties for other users of modems. This particularly applies to bulletin board system operators who are open to pressure from telephone authorities, as well as having to put up with a great number of questions from interested journalists.

CHAPTER 18
Coming Attractions

What has been described in this book is currently available and sufficiently well understood for amateurs to operate and organize. There are a number of developments already in use for professionals which will become publicly available over the next few years. The dominating factor will probably be that of cost. Until the transmission of data is generally cheaper than it is today many of these developments will not be successful in Britain. The telephone authorities have modified their tariffs for certain types of call, but no special rates are yet available for computer calls. When (as it surely must) this happens, there will probably be an enormous surge forward.

There are already some reduced rate call methods available to users. In the United States organizations such as MTI and Sprint offer reduced rate long-distance calls. This works by phoning them through an ordinary number and then using their own separate long-distance lines. In Britain there are two possibilities; Midnight Line and PSS. Midnight Line is a service where, by paying a large quarterly fee, a user may make an unlimited number of calls within Britain after midnight. One day this may be extended to cover other countries. The most attractive proposition is the PSS network. This network may be called with an ordinary modem, but the network itself works in a different manner. You may only call other users and networks connected to it. At the present time there is little of interest to the amateur apart from Compuserve and The Source in the United States

The most enjoyable of the newer ideas is that of on-line multi-player games, in which a number of separate players compete or join to play giant games. It is quite a shock to play a computer game and to have to worry about cheating or being watched by other players. In multi-player games the computer only sets the scene. You don't just play against a computer, but you play with other humans. Most of the other players will try to help you, but not all. You will never be sure who is using a false name to trap you. A multi-user dungeon commonly known as MUD has been running experimentally in England for several years and is played with great enthusiasm by many addicts. The game is expected to be available commercially in 1985 from Century Communications Ltd.

As computing becomes more powerful it is likely that home users will be offered facilities to use large mainframe computers of the types used by business. The first signs of this have come with Prestel where a gateway is possible which goes out of Prestel and into some other computers. Home banking is being offered by one building society and one bank and this is sure to spread. Home banking makes it possible to check one's bank balance at any time from home, look at a bank statement and pay bills. This will save a great deal of work for the banks eventually and will probably become very popular. When predicting future uses, it is worth remembering that Spreadsheets were unknown just a few years ago. Now spreadsheets are the most common piece of software used by businesses. It isn't likely that people will start using large computers for their personal diaries, or as name and address books. It will probably stay more convenient to use pencil and paper for most of these jobs for years to come. The advances will come in small unexpected areas. We may see advances whereby a central computer looks after a house, and starts cooking while the residents are away, but, personally, I feel that such ideas will never become widespread as they offer little advance on simpler methods such as time-clocks.

Message handling will surely develop over the next few years. Already the commercial systems are being linked up very efficiently. You can call Prestel and send a telex to a non-Prestel subscriber. It is easy to call a foreign data network. Many of the message systems will allow a message to be entered in one system and transferred to the destination user's system quite automatically, but, so far, amateur communications and bulletin boards have not offered this facility. Bulletin boards aren't likely to develop this kind of service and compete with commercial systems. They will become more localized in the future.

With the availability of home to mainframe computers, more work is likely to be done from home. At first, this is likely to be ordinary typing for typesetters or general secretarial work, but accounting and administration work may follow. It is probably an exaggeration to suggest that everyone will work from home one day, but by reorganizing work more time will certainly be spent at home. A pattern of meetings and general business in the morning followed by dicatating, typing or general computer work from homes in the afternoon can be imagined. How widespread this becomes will depend on the economics.

Shopping from home by computer is available now, but seems to offer few advantages to the consumer. The progress here will probably come in the area of stockbroking. At the moment stockbrokers cannot economically handle small customers and clients but with the aid of computers and callers from home, it would be possible to offer an on-line broking service. The attraction in allowing orders for shares to be placed by computer, avoiding all the expensive clerical work by skilled clerks, will tempt a few people to try this experiment before too long.

I have stressed the economic point in all these possibilities because it is the factor which determines whether a given option really survives. Only a few individuals are prepared to offer free public bulletin board services and consequently the future will probably depend on commercial suppliers. Other factors may also become important. For example, in France the telephone directories are now being replaced by an on-line information system which may be called direct; this saves the enormous cost of compiling, printing and distributing telephone directories to everyone every year.

For Atari Users

A. Starting Up

JTERM is a public-domain download/upload terminal program for the ATARI computer written by Frank Jones. Connect your modem to port #1 of the 850 interface module. Load the BASIC cartridge into your computer and, if you are using a disk, be sure that the TYPE 850 DISK DRIVE is turned on and that the disk in drive D1 contains an AUTORUN.SYS file which boots the RS-232 handler. The DOS and NEW commands wipe out the RS-232 handler, so you must re-boot the handler after using either. Turn on your computer, and with a disk containing JTERM in your disk drive, RUN 'D:JTERM'. JTERM begins operation by entering a machine language routine into a string; while this is going on the screen will be black. The first menu will then appear on the screen along with information about the size and location in RAM of the available text buffer. You make all menu choices by typing the appropriate key that is highlighted in inverse video. (Type an ordinary, not an inverse video character.)

In the first menu JTERM asks you to select whether you wish to download a file from a remote computer or upload a file to that computer. If you will be doing simple communication without file transfer, select download. The second menu asks you to select the translation mode: None, Light, or ATASCII. If you select Light, the 850 sets the high-order bit of each outgoing and incoming character to 0, and changes the ATASCII EOL character (155) to the ASCII CR character (13) on output and vice versa on input. If you pick None, the 850 changes no characters during either input or output. However, JTERM does some translation itself--see Section C. In ATASCII mode, neither the program nor the 850 does any translation on output. For communication with non-ATARI computers the correct choice is usually Light.

The next menu has you select outgoing parity: None, Odd, Even, or Set. Pick None if you chose no translation or ATASCII mode, since any other choice can change the high-order bits you wanted to preserve. Otherwise, follow the specific parity requirement of the computer to which you will be connecting. Note that incoming parity is not checked or changed by JTERM.

At this point, if you selected upload mode, JTERM will ask you for the filespec of the file you want to upload. When you enter the filespec, JTERM will load the file into RAM and then list it to the screen. JTERM will then enter the terminal mode. If you selected download mode, JTERM will enter terminal mode without going through the file loading routine.

B. Terminal Operations

When JTERM enters the terminal mode, the flag (inverse video word) TERMINAL appears at the top of the screen. You can now call the remote computer of your choice and log on. (The Hayes Smartmodem cannot be dialled from the keyboard if JTERM is in ATASCII mode.) Terminal mode is the mode in which you communicate interactively with the remote computer. Press the ⟨SELECT⟩ button to toggle the memory save function on and off; the flag MEMSTORE ON or MEMSTORE OFF is displayed on the screen as appropriate. When the memory-save option is on, incoming characters are stored in sequence until the buffer is full. If the buffer fills up the flag MEMORY FULL is displayed.

If you are in upload mode, do not turn on the memory-save feature until you have transmitted the contents of the buffer to the remote computer. If you do, the incoming characters will overwrite the contents of the buffer.

JTERM enters terminal mode in full duplex; that is, only those characters received from the remote computer are printed on the screen and stored in memory. If the remote computer operates in full duplex (echoes all characters that it receives back to you) the characters you send will be displayed on your screen. If the host computer operates in half duplex (does not echo the characters it receives back to you) you should turn on the half duplex mode to view your transmissions. You can toggle between half and full duplex by pressing the ⟨OPTION⟩ button. When you press ⟨OPTION⟩, the flag HALF DUPLEX or FULL DUPLEX will be displayed on the screen as appropriate.

C. Leaving Terminal Mode

To leave terminal mode press the ⟨START⟩ button. One of three things will then happen, depending on the circumstances:

(1) If JTERM is in upload mode and you have not uploaded the file, the flag UPLOADING will appear on the screen and JTERM will send the file in the buffer to the remote computer. If the remove computer echoes your transmissions, JTERM will display them on your screen. When the upload is complete JTERM reenters terminal mode in download mode. To stop the upload process before transmission of the buffer file is completed, press and hold the ⟨OPTION⟩ button until the terminal flag is displayed.
(2) If JTERM is in download mode, and you saved nothing to the memory buffer, you will return immediately to the main menu, and may start another session with different parameters if you wish.
(3) If JTERM is in download mode and you have saved something to the memory buffer, you will be asked for the filespec of the device to which you wish to save it. This can be a disk file (D:FILENAME) or the cassette (C:), printer (P:), or screen editor (E:). On entering a filespec, the contents of the memory buffer are written to the file specified. Pressing ⟨START⟩ then will enter terminal mode, or you can save the memory to a second file, on returning to terminal mode press ⟨STA ⟨OPTION⟩. This will bring you back to the request for a filespec. You can repeat this indefinitely. Rather than entering a filespec when requested, you can hit ⟨RETURN⟩ to display three other options: Push the ⟨START⟩ button to erase the buffer and return to the mode-selection menus; ⟨OPTION⟩ to erase the buffer and enter terminal mode; or ⟨SELECT⟩ to retain the contents of the buffer and return to the menus.

D. Internal Translations and Other Features.
The selection of translation modes in the second menu determines the translations made by the 850 Interface Module. Additionally, JTERM does some translation of its own. However, JTERM changes nothing it receives from the port before storing it in memory. Therefore in no translation or ATASCII mode JTERM will save in memory everything EXACTLY as it was sent. JTERM does some translation, however, before it displays to the screen.

JTERM does not display control characters (those with ASCII values less than 32) on the screen. This means that you will not see line feeds, for instance. They will, however, be stored, and could mess up a program that you are downloading; you do not need line feeds. The ASCII carriage return character (13) is translated to the ATASCII ⟨RETURN⟩ or EOL character (155). The ASCII printer bell character (7) is translated to the ATASCII console bell (253). Finally the ASCII backspace character (8) is changed to the ATASCII DELETE/BACKSPACE (126). Again, none of this translation affects what is stored in memory; everything is stored exactly as it is received.

On output in light translation or no translation mode, two characters are changed. The ATASCII DELETE/BACKSPACE character is changed to the ASCII backspace character so that the key will have the same function with most host computers that it does in the ATARI. Also the ATASCII ⟨RETURN⟩ or EOL is changed to the ASCII carriage return before it is sent. In ATASCII mode, neither of these translations is made, and the data transmitted is the ATASCII value of the character sent.

An additional feature of JTERM is the ability to send a computer 'break' by simply pressing the ⟨BREAK⟩ key. This will cause the screen to flash, a beep to sound, the flag BREAK to be printed on the screen, and last (but not least) a true break signal (approx. 0.5 sec. of SPACE tone) to be sent. The memory-save feature will be turned off and the duplex will be set to full; both should be reset to the desired modes. Do not press ⟨BREAK⟩ more than once in quick succession; if you do the second press may stop operation of JTERM. Sending the break signal will not be of much use when connected to a BBS since most of them do not recognize it, but it can be essential when you are connected to a mainframe computer whose attention cannot be got any other way.

If you press ⟨BREAK⟩ and none of the described events happen, your ATARI 850 interface has the revised RS-232 handler (there seems to be no other way to tell). In that case, remove the REM in front of the statements on line 90. Do not remove the REM if the ⟨BREAK⟩ key works correctly with it in.

```
10 REM JTERM   (VERSION 3.5)                  by Frank C. Jones Oct. 23, 1982
20 DIM PROG$(379),PROG2$(7),SPOOL$(17),IN$(26)
30 CON=53279:POKE 559,0:POKE 203,128:POKE 204,0
40 FOR I=1 TO 379:READ A:PROG$(I,I)=CHR$(A):NEXT I
50 DIM MSG$(65):RESTORE 2000:FOR I=1 TO 65:READ A:MSG$(I,I)=CHR$(A):NEXT I
60 DIM S$(5),T$(8),U$(9):FOR I=1 TO 5:READ A:S$(I,I)=CHR$(A):NEXT I:FOR I=1 TO 8
:READ A:T$(I,I)=CHR$(A):NEXT I
70 FOR I=1 TO 9:READ A:U$(I,I)=CHR$(A):NEXT I:DIM BR$(7):FOR I=1 TO 7:READ A:BR$
(I,I)=CHR$(A):NEXT I
80 FOR I=1 TO 7:READ A:PROG2$(I,I)=CHR$(A):NEXT I:FLAG=0
90 IF PEEK(8448)=182 THEN FOR I=1 TO 3:READ A:POKE 8457+I,A:NEXT I:FOR I=1 TO 8:
READ A:POKE 1663+I,A:NEXT I
95 GOSUB 2080:N=FRE(0)-256:DIM TXT$(N)
100 SETCOLOR 2,9,0:PROG$(200,200)=CHR$(13):PROG$(192,192)=CHR$(8)
110 POKE 82,0:PRINT "}";
120 PRINT N-1;" BYTES OF MEMORY AVAILABLE":PRINT "FROM-";ADR(TXT$);" TO-";ADR(TX
T$)+N-2
130 CLOSE #1:OPEN #1,4,0,"K"
140 POKE 752,1:PRINT "..Operation Mode:":PRINT :PRINT "";CHR$(196);"ownload":PRI
NT :PRINT "";CHR$(213);"pload"
145 PRINT ".(or 1-4 for disk file menu.)"
150 POKE 559,34:POKE 752,0:GET #1,ANS:IF ANS=68 THEN UPLD=0:GOTO 180
160 IF ANS=85 THEN UPLD=1:GOTO 180
165 IF ANS>48 AND ANS<53 THEN TRAP 110:QQ=ANS-48:GOSUB 760
170 GOTO 110
180 POKE 752,1:PRINT "}..Translation Mode:":PRINT :PRINT "";CHR$(206);"one":PRIN
T :PRINT "";CHR$(204);"ight"
190 PRINT :PRINT "";CHR$(193);"TASCII"
200 POKE 752,0:GET #1,ANS:IF ANS=76 THEN MODE=0:GOTO 240
```

```
210 IF ANS=78 THEN MODE=32:GOTO 240
220 IF ANS=65 THEN MODE=32:PROG$(200,200)=CHR$(155):PROG$(192,192)=CHR$(126):GOT
O 240
230 GOTO 180
240 POKE 752,1:PRINT "}..Parity:":PRINT :PRINT "";CHR$(206);"one":PRINT :PRINT "
";CHR$(207);"dd"
250 PRINT :PRINT "";CHR$(197);"ven":PRINT :PRINT "";CHR$(211);"et"
260 POKE 752,0:GET #1,ANS:IF ANS=78 THEN PARITY=0:GOTO 310
270 IF ANS=79 THEN PARITY=1:GOTO 310
280 IF ANS=69 THEN PARITY=2:GOTO 310
290 IF ANS=83 THEN PARITY=3:GOTO 310
300 GOTO 240
310 IF UPLD THEN GOSUB 490
320 PRINT "}";T$:POKE 65,0:IF  NOT FLAG THEN A=ADR(TXT$)
330 CLOSE #2:OPEN #2,13,0,"R":XIO 38,#2,MODE+PARITY,0,"R":XIO 34,#2,192,0,"R":XI
O 40,#2,0,0,"R"
340 A=USR(ADR(PROG$),A,ADR(TXT$)+N-1,ADR(MSG$)):IF PEEK(207)=128 THEN 590
350 IF A=ADR(TXT$) AND  NOT UPLD THEN CLOSE #2:GOTO 100
360 ON UPLD+1 GOSUB 430,650
370 IF UPLD THEN UPLD=0:TXT$="":GOTO 320
380 PRINT "PRESS ";S$;" TO RE-ENTER TERMINAL MODE"
390 IF PEEK(CON)<>6 THEN 390
400 IF PEEK(CON)=6 THEN 400
410 IF PEEK(CON)=2 THEN 360
420 GOTO 320
430 CLOSE #2:? "}....ENTER OUTPUT FILENAME":? "1-4 FOR DISK FILE MENU":? "     OR
 HIT <RETURN> FOR OPTIONS":? :? "";
435 POKE 702,64:POKE 65,3:TRAP 560:INPUT SPOOL$:FLAG=0:IF SPOOL$<>"" THEN 470
```

```
440 ? "}..<START> erases buffer; to menus":? "<SELECT> retains buffer; to menus"
:? "<OPTION> erases buffer; to terminal"
445 I=PEEK(CON):IF I=5 THEN FLAG=1:GOTO 100
450 IF I=6 THEN 100
460 IF I=3 THEN 320
465 GOTO 445
470 TRAP 475:QQ=VAL(SPOOL$):IF QQ>0 AND QQ<5 THEN GOSUB 760:GOTO 430
475 TRAP 430:CLOSE #3:OPEN #3,8,0,SPOOL$:IF SPOOL$(1,1)="E" THEN SETCOLOR 2,9,0
480 TXT$(A-ADR(TXT$)+1)=" ":PRINT #3;TXT$:CLOSE #3:RETURN
490 PRINT "}...ENTER UPLOAD FILENAME":? "OR 1-4 FOR DISK FILE MENU":PRINT :PRINT
 "";:POKE 702,64:INPUT SPOOL$:TXT$=""
495 TRAP 500:QQ=VAL(SPOOL$):IF QQ>0 AND QQ<5 THEN TRAP 490:GOSUB 760:GOTO 490
500 TRAP 560:CLOSE #3:OPEN #3,4,0,SPOOL$:TRAP 4:POKE 65,3
510 AD=ADR(TXT$):XX=INT(AD/256):WW=AD-XX*256:ZZ=INT((N-1)/256):YY=(N-1)-ZZ*256
520 IOCB=3:GOSUB 620:TXT$(QQ+1)=""
530 IF PEEK(883)=136 THEN 550
540 PRINT "ERROR ";PEEK(883);" DURING TEXT LOAD":STOP
550 CLOSE #3:PRINT TXT$:FOR I=1 TO 500:NEXT I:POKE 203,255:RETURN
560 PRINT "}....UNABLE TO OPEN ";SPOOL$:PRINT "PRESS ";S$;" WHEN READY"
570 IF PEEK(CON)<>6 THEN 570
580 GOTO PEEK(186)+256*PEEK(187)-10
590 CLOSE #2:SETCOLOR 2,13,10:SOUND 0,30,10,15:XIO 34,#2,2,15,"R":FOR I=1 TO 20:
NEXT I:XIO 34,#2,3,0,"R"
600 SOUND 0,0,0,0:SETCOLOR 2,9,0
610 PRINT BR$:GOTO 330
620 POKE 834+IOCB*16,7:POKE 836+IOCB*16,WW:POKE 837+IOCB*16,XX:POKE 840+IOCB*16,
YY:POKE 841+IOCB*16,ZZ
630 K=USR(ADR(PROG2$),IOCB*16)
640 QQ=PEEK(840+IOCB*16)+256*PEEK(841+IOCB*16):RETURN
```

```
650 PRINT "}....";U$
660 LL=LEN(TXT$):LN=INT(LL/25)+1
670 FOR I=1 TO LN
680 IF I=LN THEN PRINT #2;TXT$((I-1)*25+1);:GOTO 700
690 PRINT #2;TXT$((I-1)*25+1,I*25);
700 STATUS #2,B:BY=PEEK(747):IF BY THEN GET #2,A:PRINT CHR$(A);:GOTO 700
705 IF PEEK(CON)=3 THEN POP :POKE 203,128:RETURN
710 NEXT I
720 FOR I=1 TO 20
730 STATUS #2,B:BY=PEEK(747):IF BY THEN GET #2,A:PRINT CHR$(A);:GOTO 730
740 NEXT I
750 PRINT "UPLOAD COMPLETE":FOR I=1 TO 500:NEXT I:POKE 203,128:RETURN
760 SPOOL$="D?:*.*":SPOOL$(2,2)=STR$(QQ):POKE 65,3:PRINT CHR$(125):CLOSE #5:OPEN
 #5,6,0,SPOOL$:TRAP 780
770 INPUT #5,SPOOL$:PRINT SPOOL$:GOTO 770
780 CLOSE #5:PRINT "PRESS ";S$;" TO CONTINUE"
790 IF PEEK(CON)=6 THEN RETURN
800 GOTO 790
1000 DATA 104,104,133,213,104,133,212,104,133,215,104,133,214,104,133,225,104,13
3,224,169,128,234,234,169,0
1010 DATA 234,234,133,207,172,31,208,192,7,240,112,192,6,208,1,96,192,5,208,32,1
72,31,208,192,5
1020 DATA 240,249,164,203,192,255,240,90,152,73,128,133,203,208,6,169,12,133,217
,208,36,169,25,133,217
1030 DATA 208,30,192,3,208,67,172,31,208,192,3,240,249,164,204,152,73,128,133,20
4,208,6,169,51,133
1040 DATA 217,208,4,169,38,133,217,24,165,224,101,217,141,68,3,165,225,105,0,141
,69,3,169,14,141
```

```
1050 DATA 72,3,169,0,141,73,3,169,11,141,66,3,162,0,32,86,228,169,0,240,2,240,13
7,173,252
1060 DATA 2,201,255,240,54,162,32,169,11,157,66,3,169,0,157,72,3,157,73,3,162,16
,157,72,3
1070 DATA 157,73,3,169,7,157,66,3,32,86,228,201,126,208,4,169,8,208,6,201,155,20
8,2,169,13
1080 DATA 162,32,32,86,228,164,204,208,50,165,17,208,9,169,128,133,17,133,207,96
,240,243,162,32,169
1090 DATA 13,157,66,3,32,86,228,173,235,2,201,0,240,163,169,7,157,66,3,169,0,157
,72,3,157
1100 DATA 73,3,32,86,228,192,154,240,210,164,203,208,10,162,0,129,212,230,212,20
8,2,230,213,201,13
1110 DATA 208,4,169,155,208,20,201,7,208,4,169,253,208,12,201,8,208,4,169,126,20
8,4,201,32,144
1120 DATA 18,160,11,140,66,3,160,0,140,72,3,140,73,3,162,0,32,86,228,165,215,197
,213,144,16
1130 DATA 240,2,208,136,165,214,197,212,144,6,240,4,169,0,240,135,169,255,133,20
3,165,224,141,68,3
1140 DATA 165,225,141,69,3,169,13,141,72,3,169,0,141,73,3,169,11,141,66,3,162,0,
32,86,228
1150 DATA 169,0,240,216
2000 DATA 155,205,197,205,207,210,217,160,198,213,204,204,155,205,197,205,211,21
2,207,210
2010 DATA 197,160,207,206,160,155,205,197,205,211
2020 DATA 212,207,210,197,160,207,198,198,155,200,193,204,198,160,196,213,208,20
4,197,216,160,155
2030 DATA 198,213,204,204,160,196,213,208,204,197,216,160,155
2040 DATA 211,212,193,210,212
2050 DATA 212,197,210,205,201,206,193,204
```

```
2060 DATA 213,208,204,207,193,196,201,206,199,155,194,210,197,193,203,155,104,10
4,104,170,76,86,228
2070 DATA 32,128,6,141,14,210,169,0,133,17,96
2080 POKE 842,13:? "}":POSITION 2,6:FOR I=1000 TO 1150 STEP 10:? I:NEXT I:? "CON
T":POSITION 0,0:STOP :LIST 100,200
2090 ? "}":POSITION 2,6:FOR I=10 TO 90 STEP 10:? I:NEXT I:? "CONT":POSITION 0,0:
STOP :LIST 100,200
2100 ? "}":POSITION 2,6:FOR I=2000 TO 2100 STEP 10:? I:NEXT I:? "G.2110":POSITIO
N 0,0:STOP
2110 POKE 842,12:RETURN
```

APPENDIX A

Simple Terminal Programs

This is a short program that will allow a BBC microcomputer to call into a bulletin board at 300 Baud. The program is very simple and is only intended to get you started. It does, however, have the ability to download (or collect) another program written in BASIC. This means that, once you have typed this program in, you can then use it to collect a more advanced program from wherever you are calling. The program was written by D.R. Boyd who has made other terminal programs freely available for BBC users.

* BBC Downloading Terminal by D.R.Boyd *

Instructions:

RUNing the program will put you in terminal mode ready to use the BBS. Pressing BREAK at any time will return you to this mode.

To Download:

(1) Instruct the BBS to download the required file in ASCII (or text) format.
IMPORTANT: Arrange this so that all that is required to start the sending of the file is one press of RETURN.
(2) Press f0. You should now see the file being transferred as though it were being typed at the keyboard.
(3) When the transfer is complete, or if you made a mistake press BREAK. You may now finish and log off the BBS, or try again from step (1).

(4) Pressing f1 will select the downloaded program, which may be SAVEd as normal.

In most cases the terminal and the downloaded program will happily reside in memory at the same time and be RUN independently. Hence you can LIST and RUN the downloaded software before logging out to make sure it's OK.

```
  0 CLS
  1PRINT"*BBC DOWNLOAD PROGRAM V2 * BY D.R. BOYD *"
  2 PRINT "f0 - Send RETURN then download"
  3 PRINT "BREAK - Reselect Terminal Mode."
  4PRINT "f1 - Select Downloaded File."
100D%=(TOP DIV256+2)*256
110 *KEY0|@PAGE=D%|MNEW|M*FX138,2,13|M*FX2,1|M
120*KEY1|@PAGE=D%|M
130 *KEY10OLD|MRUN|M
140ONERROR GOTO250
150*FX7,3
160*FX8,3
170REPEAT
180*FX2,1
190*FX3
200B%=INKEY(0):IFB%>31 OR B%=10 OR B%=13 VDUB%
210 *FX2,2
220 *FX3,7
230 B%=INKEY(0):IFB%>0 VDUB%
240 UNTILB%=0
250 *FX2
260 *FX3
```

APPENDIX B

AFPAS Standards 1983

Standards and Protocols for Free Public Access Bulletin Boards Published by The Association of Free Public Access Systems

1. These standards were agreed by the Association of Free Public Access Systems (AFPAS) in March 1983.

2. The standards will be implemented on Public Bulletin Boards connected to the Public Switched Telephone Network.

3. Frequencies and Baud Rates to be used.

3.a. The tones used will be to CCITT Standards using 300 Baud (V.21) definitions.

3.b. A secondary option of 1200 baud transmission and 75 Baud reception (V.23) may optionally also be used.

3.c. Bell System tones will not generally be made available except as a late night option after 00:00 hours and before 09:00 hours.

4. Information will be sent using ASCII encoding of all text for display

5. Graphics whether to Prestel/Teletext standards or to NAFLPS standards shall not be used unless confined to a separate area of the bulletin board.

6. Serial Word Lengths Allowable.

6.a. All boards shall be able to accept calls in a seven data-bit word format with:

1 Start-Bit + 7 Data-Bits + 1 Even Parity-Bit + 1 Stop-Bit

6.b. To facilitate the 'Christensen Modem Protocols' boards may also be run in an eight data-bit format with no parity bit present. The conditions in 6.a. shall still apply.

6.c. Parity bits may be generated, but should not be checked on the reception of text when working as described in 6.a.

7. Certain non-text control codes shall have special purposes as follows:-

7.a. X-On / X-Off

7.a.i. On the receipt of an X-Off signal (Control-S, DC3 or Hexadecimal 13) the receiving host bulletin board shall immediately cease transmission of characters after the completion of the current character (10 serial bits).

7.a.ii. On the receipt of an X-On signal (Control-Q, DC1 or Hexadecimal 11) the transmission shall be continued again from the point at which it ceased.

7.a.iii. Optionally the system may allow continuation of text after the receipt of any character. This is to facilitate naive user use.

7.b. Automatic Log-On

7.b.i. The host system may optionally allow automatic user log-on. The host system will initiate this by sending a single Control-E (Decimal 5).

7.b.ii. The caller shall reply with an upper case ASCII string of the following form :-

FIRST-NAME;LAST-NAME;LOCATION⟨Carriage-Return⟩

7.b.iii. If the caller does not reply in a string as above then the system will prompt the caller for a manual log-on procedure.

7.c. Buffer Control

7.c.i. To facilitate the accurate reception of data-files the board may send a signal to the incoming caller instructing the caller's terminal to open an empty buffer area in memory and to accept all characters in a continuous stream into that buffer. In order to open the buffer the code Control-R shall be sent.

7.c.ii. On receipt of a Control-T code the buffer shall be closed and no more characters added to it.

7.c.iii. The user should have some way of storing the buffer contents to a permanent form of storage.

7.c.iv. The buffer open and close commands should not themselves be stored in the buffer.

7.d. In case of a wrongly typed character Control-H (Hexadecimal 8) shall be sent to act as a back-space character.

8. Data File Transmissions

8.a. ASCII text data files may be sent by either method 8.c.i. or 8.c.3. detailed below.

8.b. Binary data files (i.e. those containing any non-printable ASCII characters) may be sent by either method 8.c.ii or 8.c.iii.

8.c. There are three acceptable methods for file transmission.

8.c.i. Pure ASCII text optionally used with the buffer control codes described in Paragraph 7.c.

8.c.ii. As expanded hexadecimal ASCII for example Binary 10011100 would be sent as two ASCII text characters '9' and 'C'.

8.c.iii. By using the 'Modem' series of protocols for the transmission of data in 128 byte check-summed blocks as devised by Christensen. This protocol includes full handshaking and re-transmission of blocks incorrectly received. The software and algorithms are Public Domain Programs and may be obtained from the CP/M User Group.

APPENDIX C

U.K. Free Public Access Bulletin Boards

These boards are run mainly by private individuals and so the numbers may change from time to time with the operator's personal circumstances. Other numbers for bulletin boards can be obtained from lists published by various monthly magazines.

FORUM-80 Hull
Telephone 0482-859169 - 300 Baud Only
Tuesdays & Thursdays 19:00 to 22:00
Saturdays & Sundays 13:00 to 22:00
Daily (USA Bell 103) 00:00 to 08:00

TBBS London Local
Telephone 01-348-9400 – 300 Baud Only
24 Hours Every Day

TBBS Blandford
Telephone 0258-54494 – 300 Baud Only
24 Hours Every Day (Dorset Area)

TBBS Liverpool – Mailbox 80
Telephone 051428-8924 – 300 Baud Only
24 Hours Every Day

CABBS London
Telephone 01-631-3076 – 300 and 1200/75 Baud
24 Hours Every Day

Southern Bulletin Board
Telephone 0243-511077
24 Hours Every Day (Chichester Area)

Mailbox 83
Telephone 0384-635336 – 300 Baud RINGBACK
17:30 to 08:30 Monday – Saturday
24 Hours on Sunday (West Midlands Area)

CBBS South West
Telephone 0626-890014 – 300 Baud
24 Hours Daily

CBBS Surrey
Telephone 04862-25174 – 300 Baud

CBBS Chiltern
Telephone 07073-28723 – 300 Baud
18:30 to 22:30 Daily